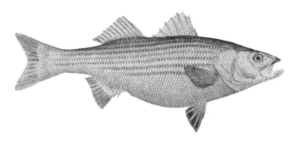

"That fish, that is not catched thereby,
Alas, is wiser far than I"

- John Donne 17th century British poet

Barnegat Bay's Fisheries

Richard A. Henderson, Jr.

Blue Claw Books

Barnegat Bay's Fisheries

First Printing: June 1997.

Cover Art by Jim Doucette. Based upon the color plate from McClane's Field Guide to Saltwater Fishes of North America.
Illustrations on pages 1, 8 and 9 courtesy of the Smithsonian Institute, Washington, D.C.
Maps on pages 24, 25, 46, 47, 64, 65, 84, and 85: Charterboat Charts Inc.
Cover design, diagrams, layout, photography (excluding pages 30, 53, 61, 71, 75, 93, 124, 137, 140, 154, 157), and all other illustrations by Richard A. Henderson, Jr.

Special Thanks

Chris who often shared the trips that wrote this book.
Jim whose unique brand of humor often turned serious trips into encounters with zomeeses.
Debbie, Loretta and Jim who honed my writing skills through papers, papers and more papers! I owe them much.
Manny Luftglass for his kind advice.

First Edition

ISBN 0-9658470-0-4

Library of Congress Catalog Card Number: 97-72746

To my mother, father and brother who share my love of fishing.

To Janet for her loving support of this book and me.

An Angle on Barnegat Bay

Whether fishing to you is waging an epic battle with a striped bass amid the fury of a crashing surf or fishing for snapper bluefish in a tepid marsh, I am confident your curiosity is aroused when you see literature discussing a favorite fishing area. Writings about a favored fishing location make for great reading not because they're filled with practical information or funny imagery, but because you can relate to what the writer is discussing through his or her words.

As a young boy, I often scoured magazines and books for any scrap of information relating to the fishing I was experiencing at Barnegat Bay. Far away, exotic places where some angler went head to head with a half ton marlin were often covered in intricate detail by most fishing magazines, but stories discussing Barnegat Bay's great fishing or crabbing were like water in the desert — scarce.

Therefore, I thought a book that examined Barnegat Bay from an angling and crabbing standpoint — the species that swim its waters, where to find them and what tackle and techniques are best suited for coaxing a few to put a bend in your rod — would be a good idea.

If you're an old salt who has weathered many years on the bay, this book may give you a few ideas about a new bait or location to fish if your favorite bait or location bombs. For the vacationing angler who's looking to try some fishing before heading home, prepare to be introduced to the bay's favorite angling species with advice on how to make an initial trip a positive experience.

First-hand observation will serve an angler best when fishing, and this book's intent is to heighten that observation

Richard A. Henderson, Jr.

Contents

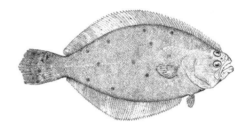

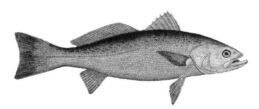

Tog & Sea Bass   Page 91

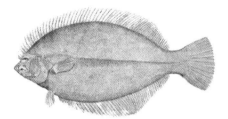

Winter Flounder   Page 107

Blowfish Etcetera   Page 125

Blue Claw Crabs   Page 143

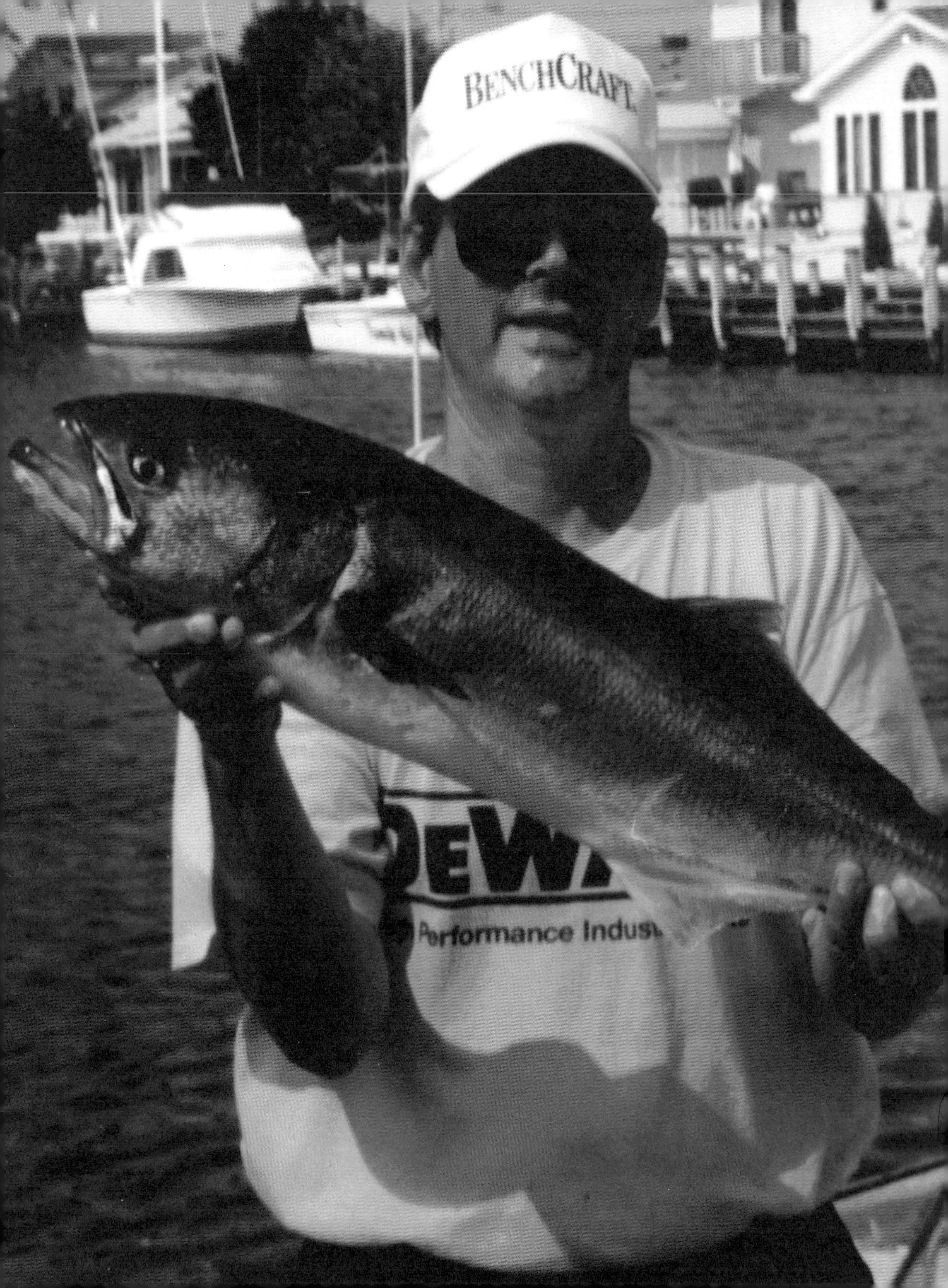
BENCHCRAFT
DeW
Performance Indus

Chapter 1
Bluefish

Few saltwater anglers can resist the excitement bluefish offer – the sight of mullet scattering like raindrops across the surface of a bluefish-induced frothing sea while sea gulls pick at the frantic baitfish. Bluefish, scientifically known as Pomatomus saltatrix, provide great sport for anglers whether they are in the guise of 6-inch "snappers" or 15-pound "alligators."

Bluefish are an attractive looking fish. Their head and upper body above the pectoral fin range in color from a dark navy to a brilliant greenish-bronze that quickly fades to a silvery white or very light gray along the bluefish's flanks and stomach. Equipped with a tooth-filled mouth and a strong caudal fin (tail), a bluefish is built for chasing down and seizing baitfish in the open water.

Bluefish travel together in schools, which is good for anglers because once one locates a single bluefish, there are many more within close proximity. Bluefish schools are usually comprised of similar sized fish, mostly because larger bluefish have a tendency to eat their smaller brethren as if they were a bunker or mullet.

Juvenile bluefish, commonly referred to as "snappers" or "snapper blues," are in the 3 to 10-inch range and are one of Barnegat Bay's more conspicuous fish. Snappers thrive within the bay's creeks, lagoons, rivers and shallow water "flats" because these locations offer them protection from predators plus thick baitfish schools to dine upon. In addition to providing great light tackle sport, snappers make dynamite fluke and weakfish bait.

"Tailor" or "cocktail" blues average around 1 to 5-pounds and are the predominant sized bluefish anglers are likely to encounter in the bay, especially during May and June. Tailor bluefish love to patrol the bay's flats, channels and sand bars for baitfish schools, frequently causing the water's surface to erupt with their feeding. Tailor blues are just as aggressive and unbridled as snappers making them fun to catch with live bait, plugs, poppers, and jigs. Besides the great sport tailors provide, many anglers target these size bluefish because they make excellent table fare. Their meat is not as oily and "strong flavored" as the flesh of a large "alligator" bluefish. Plus, scientific

Barnegat Bay plays host to "alligator" blues each summer. Anglers fishing the inlet have a shot at these large bluefish (opposite).

studies conclude, at least for now, that smaller sized bluefish have less PCB contaminants in their flesh than larger bluefish making them safer to consume (we hope).

Anglers nickname large bluefish in the 10-pound plus range "alligator," "gorilla," or "slammer." Anglers who have caught these size bluefish agree the fish's strength and tenacity makes these nicknames very apropos. These size bluefish will give you and your rod a real workout. Be sure to watch your fingers when unhooking an alligator bluefish because these fish get their nickname from their large tooth-filled mouth and quick alligator-like "snaps."

Alligator bluefish are more commonly found in the ocean off Island Beach and Long Beach Island. They do, however, frequent Barnegat Inlet, and anglers fishing at the inlet's north or south jetties can catch alligators throughout the summer and especially during the fall. Alligators rarely venture deep into Barnegat Bay with Oyster Creek and Double Creek Channels' eastern ends being about as far west as you will likely find them.

Due to their flesh's strong, "fishy" smell and possible high PCB content, large bluefish are best released to fight again. If badly hooked and survival is dim, large bluefish make excellent crab bait due to their flesh's high oil content. Some people, on the other hand, eat these large specimens and claim they are quite good.

Anglers commonly refer to bluefish larger than tailors but smaller than alligators as "bluefish." How original.

Whatever their size, bluefish,

Did You Know?

The current all-tackle world record for bluefish stands at 31-pounds, 12-ounces, but a bluefish weighing 45-pounds was recorded off the coast of North Africa.

in addition to providing great sport, make superb bait for crabs and sharks while also being quite tasty on the grill.

Bait For Blues

Within nature, it is safe to assume that any smaller fish is fair game for a bluefish's wraith. Bluefish do, however, tend to have their preferences and favor oily, strong-smelling fish, such as bunker and mackerel.

Bunker, also known as Atlantic menhaden or mossbunker, are a bluefish staple. An extremely oily-fleshed fish that resembles a wide herring, bunker range from small 4-inch "peanut" specimens to 12-inch crab bait sized fish. Anglers employ bunker in a variety of ways when targeting bluefish. They chum with ground bunker while bunker steaks and chunks are often fished within the chum slick. Live bunker are simply livelined, usually for alligator blues.

Similar to many other predatory fish, bluefish respond to live or fresh bunker better than the frozen or rancid variety. Anglers can purchase frozen bunker at tackle stores and can occasionally acquire live bunker from private dealers. Aside from taking a chance with a tackle store, netting and snagging are the two more popular ways to acquire live bunker.

Anglers typically net peanut bunker with seine nets via the same seining technique employed for spearing. The old world-style cast nets come into play when anglers seek to net bunker near the surface of un-seinable waters, such as a bulkhead.

Perhaps the most productive

and fun way to catch live bunker is with a "bunker snagger." A bunker snagger typically takes the form of a weighted treble hook. Anglers cast the snagger at a bunker school and then quickly retrieve it through the school while sharply jerking the snagger through the water hoping its sharp hook snags a bunker mysteriously appear off New Jersey around the time the thick mackerel schools pass the area during their northern migration. Many anglers favor mackerel over bunker because mackerel possess firmer flesh which holds a hook better. Anglers liveline small 5-inch live mackerel while slicing

bunker. Once a bunker is snagged, some anglers retrieve the snagged fish and immediately place it on ice, to keep the fish's flesh fresh and firm for later use. Others simply let out line and let the wounded, snagged fish swim about, hoping that predators are watching. If bluefish are anywhere in the immediate vicinity one will usually hit the bunker, if of course a striped bass does not beat the bluefish to the punch!

     Mackerel, whether it be the Atlantic or tinker variety, is another oily-fleshed fish that bluefish love to hunt. It is no coincidence that the first

one or 2-inch wide steaks from large mackerel for bait. Anglers typically drift 5-inch long, 1/4-inch wide strips, cut from the mackerel's colorful flank, for bluefish while fashioning small 2-inch strips, from the mackerel's belly, for snapper bait. 1 or 2-inch wide steaks, cut vertically from the mackerel's flank, are great to use in a chum slick for blues.

     Mullet, butterfish, spearing, sand eels, squid strips, and killies are other commonly used bluefish baits whether anglers fish them whole or, in the case of large buttterfish, cut.

Bluefish attack a baitfish school off Barnegat Inlet's north jetty. The water's surface usually erupts when a bluefish school meets a baitfish school, easily keying anglers to the bluefish school's location (center).

Spearing, sand eels and squid strips make very effective baits for snappers. Bluefish, much to the dismay of striped bass anglers, also have quite a taste for eels.

Tackle and Techniques

The view some anglers hold of bluefish consists of a Looney Tunes-like chaotic mess of fish hitting anything in the water that moves. While there are a few episodes that can justify this conception, Barnegat Bay's bluefish population generally limits their attacks and travels stealthfully below the surface while reacting to bait and lures that either match or resemble their prey. The anglers who spend their time refining their bluefish catching techniques are typically the ones with the sore arms. Anglers catch bluefish through a variety of methods including: drifting, chumming, spot-casting, jigging, livelining, and trolling with the most commonly utilized techniques on the Barnegat Bay being drifting, chumming and spot-casting.

When drifting for bluefish, anglers typically employ either a fish-finder rig or a hi-lo rig. The fish-finder rig is the rig of choice when anglers want to drift bait near the bottom and have the best possible feel for when a bluefish takes the bait. Tie this rig the same way you would for fluke (see page 36) but employ a Carlisle or beak hook instead of an English wide gap hook.

The hi-lo rig is the other popularly used bluefish rig in the bay. Though it does not offer a fish-finder rig's sensitivity, the hi-lo rig lets one fish a bait near the bottom while another rides higher in the water column where bluefish tend to swim. Anglers typically bait the top "high" hook with an oily bluefish bait, such as a mackerel strip, while the bottom hook gets fitted with a squid strip or killie for fluke or weakfish. Please see page 36 for a more detailed description of how to tie a hi-lo rig.

When engineering rigs for bluefish, anglers should employ leader and hooks suited to handling a bluefish's powerful pulls and sharp teeth. When seeking bluefish with bait in the bay, anglers typically fashion their rigs out of 20 to 30-pound leader material. As far as hooks go, the Carlisle style hook is a popular bluefish hook because it possesses a long shank which prevents bluefish from biting through the line. The hook's long shank also allows for easier hook removal. The beak is another popular model employed for bluefish.

Hook sizes depend upon the size of the fish one intends to target. For tailor size bluefish, many anglers start with either a size 1/0 Carlisle or a 7/0 offset beak style hook. If hooks start to get swallowed, they use a larger size hook, such as a 4/0 Carlisle or a 5/0 offset beak hook.

It's a Snap

Though anglers can employ fish-finder and hi-lo rigs tied with small hooks when targeting snappers, fishing with a bobber is perhaps the best and most enjoyable way to catch snappers from a bridge, dock or bulkheading. All one needs to fish this way is a #9 or #10 Eagle Claw beak style hook, a golf ball-size bobber, a bb-size split shot, and bait. Simply snell the hook directly to your line and attach the bobber to the line about a foot above the the hook. Alter this length depending on the water's depth and where the snappers are in relation to the surface. For bait, snappers respond strongly to live, dead and cut baits. Particularly good snapper baits are squid strips, fresh spearing and fresh sand eels. Hook spearing and sand eels through the eyes while double hooking strips at their front. Anglers attach a bb-size splitshot about 5-inches before the bait if they want to increase their casting distance. Toss the bobber out and wait for the bobber to move. If it plunges beneath the water, strike back; if it moves in sharp vertical or horizontal bursts, wait for the snapper to firmly grasp the bait and move off with it.

Chum Some

One highly productive way to fish for bluefish in the bay is to employ chum. Chum is basically an attractant that emanates from an angler's position, attracts fish to it and then holds them there. Chumming is especially effective when done in locations that frequently attract bluefish, such as Tices Shoals, Oyster Creek Channel and Barnegat Inlet.

When considering an area and time to chum, certain areas, particularly the inlet, become choked with boats so it is wise to chum either at daybreak or dusk. Legally, anglers cannot anchor their boats within a channel which would block traffic. In any event, a constant boat barrage will usually scatter most fish you mange to attract to your position. Chumming early or late in the day usually affords anglers a

break from any strong winds that cause the boat to swing, breaking up the chum slick. When chumming, a moving tide is preferred because the tidal flow moves the chum to the fish. Also, bluefish are frequently on the move during a moving tide, either to or from the shallows. The likelihood one's chum slick will attract bluefish is significantly increased during a moving tide. The only problem with a moving tide is if it is too strong, such as at the inlet, the swiftly moving water will quickly disperse your chum.

The trick to successfully chumming bluefish is to go to a somewhat calm area and double anchor the boat up current from where the bluefish are likely to be holding. If bluefish occasionally smack the surface where a deep pocket rises up to a flat, for example, anchor your boat so when you set the anchor and drift back towards the pocket, you're about 20 yards up current from the flat. Double anchor the same way you would for winter flounder, with the first anchor dropped from the bow and a second anchor dropped from the stern. Please see page 116 for a more detailed description of double anchoring a boat. Assuming you are packin' decent chum, you should have bluefish in your slick quickly. Though anchoring in a channel, where you can obstruct boat traffic is illegal, anchoring your boat out of the channel but positioning it so the chum seeps into the channel is an effective and legal route.

After anchoring up, some anglers immediately chum bluefish to their backdoor with ground bunker while others employ the anchored boat as a steady platform from which they can pick the immediate area apart with

Did You Know?

The bluefish is the sole member of its family of fishes (the Pomatomidae family).

lures before settling down and chumming. Ground bunker, available in most tackle stores, is the standard bluefish chum. Some anglers go a step further, and dice up whole, fresh bunker and chum with the small chunks. These chunks are usually small so as to attract but not feed the bluefish. Others accentuate the standard bunker fare with whole, fresh sand eels or spearing which are good additives to a chum slick especially if one is chumming in the inlet area during July and August. Spanish mackerel often raid the inlet during this time and are especially fond of these small baitfish.

When chumming, you want to initially use undiluted, concentrated chum to attract bluefish to your position. Simply take a small mug or the old cat food can nailed to a stick and drop a scoopful of chum in the water. Once you start to get bites, dilute the chum with water. This will allow you to conserve your chum supply and keep the fish more interested in your bait rather than your chum.Whether you invest in a fancy chum bag or resort to the cat food can, you always want to have a steady stream of chum emanating from your boat because if the chum slick stops every time a fish is hooked or someone gets bored, you could loose all the fish in your slick. This will definitely put a damper on the day! Once attracted, bluefish will usually hang around for quite awhile provided you hold their attention with bait, chum or lures, and boat traffic does not disperse the school.

A good, basic rig to fish a chum slick starts with a #5 or #4 black barrel swivel connected to a 4 to 6-inch piece of #5 wire via a haywire twist. When fishing for bluefish, whether you are chumming in the inlet for 12-pound "gorillas" or poking around a salt marsh for snappers, you want to tie a piece of thin wire before your hook or lure. This wire prevents a bluefish's sharp teeth from cutting the line or chaffing it to the point where the bluefish's strength busts the line. Not to worry, bluefish, unlike Spanish mackerel or weakfish, are rarely leader shy. The length of the wire depends on the size of bluefish you are targeting. "Gorilla" blues usually require a piece about 6-inches long, while you can get away with a 3-inch leader for snappers.

You finish the rig by connecting a Mustad bronze offset beak style hook to the wire's other end via a haywire twist. Hooks range in size from a 3/0 to a 7/0 depending on the size of the bluefish. A 3/0 or 4/0 is good for tailor blues and overgrown snappers while a 5/0 and larger sizes can handle alligator blues. Many bluefish anglers substitute the beak style hook with a Limerick or a Carlisle style hook in sizes 3/0 to 6/0. Smaller sized Carlisle hooks, in the #5 neighborhood, #8 Limericks and small Sprout hooks are good to employ if targeting snappers.

Chief among the baits one should fish their slick with are fresh bunker chunks. Frozen bunker, which tends to soften when defrosted, is not nearly as good as freshly snagged bunker that is immediately chunked or kept fresh on ice and then chunked. Fresh bunker possesses a much firmer texture than frozen or rancid bunker, making it easier to chunk and then hook.

When chunking a bunker, keep a sharp knife to avoid mashing the bunker. From head to tail, cut the bunker into 1-inch wide steaks. Cut the steaks in half and then cut the halves in half. Some anglers prefer to fish a whole steak when targeting large bluefish. For the most part, bunker chunks will serve one well in the bay. Hook the chunk or steak twice, inserting the hook through the skin to anchor the hook. Snake the bunker up the hook's shank, so the hook is buried in the soft, oily bunker.

In addition to fresh bunker, chunked mackerel or butterfish are sound options and they tend to hold onto the hook better than bunker.

After hooking your bait, simply take a handful of bait chunks, drop them in the water and allow your baited hook to drift with the chunks. Keep your reel in free spool and ladle out line so your baited hook drifts with the chunks. When fishing a chum slick, you should work different depths to locate the fish. This only applies when chumming areas that are somewhat deep. If you are chumming in 5-feet of water, you do not need to employ weight to lower your bait because bluefish will find it in shallow water. When chumming deeper water though, you may want to employ some weight to lower your bait to where the fish may be holding. When chumming deep water locations within the bay, such as the inlet, many anglers have one bait drift weightlessly near the surface while another, fitted with a small rubber core sinker or split shot 2-feet before the rig, brings the bait down to determine whether the fish are holding deep. Ambitious anglers commonly jig 007 or A27 diamond jigs from the bottom steadily towards the surface. They work different depths until they either get a fish or reach the surface, and when they get a fish, they note the approximate depth and then weight their baited hooks to fish that particular depth.

The Jig is Up

Bluefish primarily rely upon their sight and open water speed to hunt and attack their prey. This, coupled with their rather bloodthirsty tendencies, makes them an ideal fish to target with artificial lures. Bluefish lures range from "metals," which refer to lures composed primarily of metal, to "poppers," which are long, slender lures possessing a concave front that causes the popper to splash water or "pop" when jerked and retrieved.

On Barnegat Bay, lures really lend themselves to "spot casting" which basically consists of spotting bluefish or baitfish breaking the water's surface and proceeding to cast at the commotion. While spot casting may be frowned upon by some savvy anglers, it does prove to be an effective "no-frills" approach to bluefishing especially when boat traffic, strong currents and narrow channels, which characterize many areas of Barnegat Bay, are involved.

When approaching a bluefish school with a boat, it's important not to spook the bluefish. Many times you will find boats carefully holding near a school's perimeter, working the interior with long casts, only to have some fool drive straight through the school and disperse the fish. As long as everyone stays at a casting distance from the school, following it slowly and staying along the school's perimeter, everyone will catch fish and have fun. It is when people get "ants in their pants" that fish

Did You Know?
The current New Jersey state record for bluefish stands at 24-pounds, 4-ounces and was taken at Atlantic City in 1985.

get dispersed.

When targeting bluefish with lures, it is important to match a lure's shape and size to the prevalent baitfish of the target fishing area. Decide at what speed does the lure produce the action that best imitates these baitfish. Basically, you are creating an illusion, and the better you are at imitating local baitfish, the more bluefish you will catch. Spearing, "peanut" bunker, sand eels, and mullet are a few prime bluefish prey that should be rendered in your lure collection.

When employing lures, your rod partly determines whether you will be able to work the lure properly. A 6 to 7-foot stiff-tipped, graphite rod is a good choice for maximum casting distance and sensitivity. Wide, large spooled spinning reels with a high gear ratio, such as 6:1, spooled with 8 to 12-pound test monofilament are well suited for this rod and for casting lures for bay bluefish. Stiff-tipped rods are especially good for plugs, jigs and poppers because once a bluefish grabs the lure, you can strike back and firmly set the hook.

Many anglers think retrieving any old plug quickly through a bluefish school will catch fish. This theory may hold true when bluefish are tightly schooled and shredding a hapless baitfish school. Once things cool down, however, bluefish tend to become inconspicuous and quite finicky. Though not as finicky as weakfish, bluefish will occasionally ignore plugs that do not resemble, for whatever reason, the local baitfish. Anglers who stick to plugs that resemble local baitfish in size, look and action will consistently score well with bluefish. Plugs that commonly see action on the bay include:

Bagleys, Rapalas, Zara Spooks, Bombers, Rebels, and Redfins, with the latter three being the more popular choices. Tobimaru makes a few plugs that work well for bluefish. One particular model possesses a beautiful finish and contains weight in its rear so when an angler retrieves it very slowly, the lure closely resembles a badly injured baitfish struggling near the water's surface.

Regardless of whatever plug you choose, the plug's coloration should somewhat match the coloration of a baitfish indigenous to the bay because bluefish often hunt by sight. A silver bodied, black backed plug resembles a mullet; a silver plug with a blue back resembles a herring; a slender, silver bodied, dark green backed plug resembles a large spearing, large sand eel or small snapper. Some anglers take plug coloration a step further and paint the belly of their plugs with red nail polish to create the illusion of a bleeding baitfish. Many surf anglers use the belly coloring trick, and according to some it really pays off.

Aside from color, bluefish often key onto size. If blues are feeding on 5-inch mullet, it is generally not productive to start heaving 10-inch plugs. For the bay, one generally wants to stick to plugs in the 4½-inch to 7-inch range. Five-inch Redfins, 5½-inch Rebels, 15A (4½-inch) and 16A (6-inch) are prime examples of sizes you want cast when targeting the bay's bluefish population.

If targeting snappers with plugs, you obviously want to heave a smaller size. Plugs in the 1-inch to 3-inch neighborhood, such as a 1½-inch Rebel or a Fenwick Pin's Minnow, work well when fishing for snappers. Many anglers will choose small metals over small plugs for snappers. A small metal will cast like a bullet whereas a small, fly-weight plug is often difficult to cast far. It's not too smart to try small plugs if larger bluefish may be in the immediate area. A larger than snapper size bluefish will usually either engulf

closely resemble the silvery bodied baitfish bluefish love to hammer. Their design and weight lets the angler cast great distances and work both the surface and bottom. Anglers can make most metal lures zig zag and skip along the surface, appearing like a fleeing baitfish, by retrieving the metal as soon

the plug, making hook removal a nightmare, or simply snap the line. Even if you employ wire leader, with an undersized plug, when you land a thrashing blue with one of the plug's 3 treble hooks firmly anchored in its jaw, the second hook down its throat, while the third hook flails wildly with the fish's every move, you'll wish you selected a larger plug. Sometimes though, you have to use small lures to match small baitfish. Rather than employ small plugs in areas mixed with snappers and larger blues, many anglers use small, single-hooked metals.

"Metals" refer to lures composed primarily of metal. Metals are standard bluefish lures because they

as it hits the surface. When the sun gets too high, and the water surface becomes too bright for a bluefish's eyes, anglers can let the lure sink to the bottom and work it slowly for bluefish and even weakfish or striped bass.

Barnegat Bay bluefish favor metals such as Crippled Herrings, Kastmasters and 007/A17 Ava jigs, Bridgeport jigs, $1/4$ to $1^{1}/_{2}$-inch Hopkins, Need-L-Eels, and Tony Acceta spoons — to name a few. When using most metals, particularly the Crippled Herrings, Hopkins and Need-L-Eels, retrieve the lure smoothly and slowly, letting the lure's action do the work for you. Some anglers imitate the irregular scales of a baitfish by denting their metal lures,

Bluefish are equipped with a powerful jaw filled with sharp teeth. Firmly grasp the fish behind its head and employ needlenose pliers or some other hook disgorger to safely remove the hook (center).

which are often flawless, by banging them against a hard object.

Within recent years, saltwater fly fishing has become quite popular among recreational anglers. Bluefish, because of their aggressiveness and surface feeding tendencies, are a

being followed by a larger fish, represented by the plug.

To many anglers, the predecessor to saltwater flies is the popper. A popper is meant to imitate a baitfish attempting to flee a predator by leaping from the water. Because bluefish

2 to 8-pound bluefish are Barnegat Bay's predominant size. Note the wire leader secured to the lure to prevent a bluefish from engulfing the lure and biting through the line (center).

favorite species among fly casters. Anglers typically flick saltwater flies at shallow water locations, particularly Reeves Cove and Tices Shoals. The Atlantic eel, sar-mul-mack, velvet eel, and crystal pencil are a few flies that will incur a bluefish's wrath. These saltwater flies also make very potent teasers when used in combination with a Bomber or Redfin plug. Simply tie the fly to a 4-inch piece of 20-pound stiff mono leader material. Tie the leaderd fly about 2-feet before your plug. When the two are retrieved in tandem, the fly/teaser will imitate a small baitfish

respond strongly to visual stimulation, a popper, with its splashing, can be a very potent lure assuming one works them properly and there are bluefish in the immediate area.

Poppers for general bay use range from $1/4$ to 2-ounces. In particular, the $1 1/4$-ounce Atoms, the Gibb's little neck popper, the Smoker, the Cordell pencil poppers, and the Creek Chub models are good poppers for the bay's bluefish.

Poppers intimidate many anglers because the bulk of the lure's action is supplied by the angler. One cannot usually just retrieve a popper as is and expect a lot of fireworks. With a little experimentation and patience, a popper can be an unequaled tool for catching bluefish.

Employing a stiff-tipped casting rod, similar to the one previously described, tie the popper directly to your line avoiding a swivel or wire leader. Basically, you want the popper to splash along the surface as you retrieve it back towards your position. To make the popper splash or "pop" along the surface, cast it at a potential fish-holding area. After allowing the popper to rest at the surface for a few seconds, point your rod at the popper, with the rod's tip slightly higher than the rod's butt. Give your reel two steady but fast turns and on the second turn, lift the rod's tip about 6-inches with a quick snap. The popper should splash or "pop." Continue to retrieve the popper slowly, giving the popper short jerks every 8 to 10-feet, mixing short jerks with stronger ones for more erratic action. Generally, bluefish react strongly to swiftly worked poppers but experiment with retrieval speed to determine the optimum pace.

Often unheralded by bluefish anglers, jigs are potent bluefish lures when fished with strip baits. A strip bait, when hooked once with the jig's hook so it flutters with the jig's movements, is an integral piece. Three or 4-inch strips cut from mackerel, mullet, butterfish or squid are good to use with jigs aimed at bluefish. Be sure to tailor the strip's size to the size of the blues in the area. Bluefish love to hit a jig's rear which results in the jig's strip bait getting severed. You do not, for example, want

to place a 5-inch strip on a jig when 12-inch snappers are marauding the area because they will likely nip your strip. Alternatively, some anglers garnish their jigs with pork rind strips, which are nearly indestructible and prevent bluefish from stripping your strip!

After landing or losing a bluefish, remember to give your hook or lure a good pull to test its connection knot. Usually, your lure or hook will break off either due to the stress from the previous fight or chaffing from a blue's sharp teeth. Checking your knot and line may sound like common sense but when the excitement of fishing is involved, common sense sometimes fails to enter the picture and this little trick will save you a lot of time (tying on a new hook/lure) and money (buying a new hook/lure).

In feeding frenzy conditions, it is a good idea to tie a small 6-inch piece of wire leader before your lure to prevent any teeth from cutting your line. Lure fanatics may holler that the wire will cause the lure's action to suffer, but when you are casting into a seething sea of baitfish, bluefish and birds, bluefish are not, during these particular episodes, the most finicky creature to ever grace the ocean. During quieter moments, however, you may want to forgo wire leader entirely or use very fine, light wire leader or heavy monofilament line, called shock leader. Shock leader, which is usually a foot or two of heavy mono line, prevents blues from slicing through thin casting line. Most anglers operating in the bay use 20-pound test. They tie their lure to one end and connect their line to the other end.

Where and When

Bluefish are found from Nova Scotia, Canada to Florida, with their greatest concentration being found between Cape Cod, Massachusetts and Cape Hatteras, North Carolina. A relatively warm water species, bluefish prefer water temperatures in the 66 to 72 degree neighborhood. They migrate northward up the coast during the spring and southward during the fall. Barnegat Bay's bluefish population is comprised mostly of bluefish that have broken off from the migrators.

One of Barnegat Bay's more predominant species, bluefish arrive at the bay around late April. Bluefishing is usually at its peak from mid-May to late June. These fish generally range from 2 to 8-pounds with larger fish making an occasional appearance. Once the summer gets into full swing around early July, bluefish spread out into the bay placing the burden of finding them primarily on the angler and not on a flock of sea gulls. The fishing is usually best at dawn and dusk. Bluefishing stays steady the rest of the summer until the blues depart around early October, depending on the water temperature. Anglers fishing from the inlet's jetties can catch bluefish into November by picking off the fish migrating south. Bluefish do not like the cooler water that striped bass prefer, so once the water temperature dips during the fall, bluefish head south.

Situations where various types of sea birds pick at frantic baitfish while bluefish make the surface come alive with their very presence takes much (if not all) of the guesswork out of locating bluefish and coaxing a few to take an offering. Situations like these, however, are not always the norm so it takes some knowledge of the bay to consistently catch bluefish. Generally, bluefish are found in areas that attract baitfish. Baitfish move with the tides and bluefish, in turn, move with the baitfish. Thus, it's smart to pay attention to the tides when figuring out where to find bluefish. Sometimes though, it's possible to find bluefish in every conceivable location in the bay save for terra firma, so detailed analysis of the moon and tides in relation to the Earth's alignment with Venus is not necessary.

Upper Bay

Located at Barnegat Bay's northern extreme is the Point Pleasant Canal, which is a man-made waterway that connects Barnegat Bay to the Manasquan River. The canal's mouth is a good area to investigate because bluefish usually follow the baitfish that use the canal to venture into the bay. Anglers fishing at the canal's mouth have a good shot at picking bluefish off as they go through the canal. Especially during the late spring, bluefish "ride the tide," entering the bay through the canal on an incoming tide with the baitfish and then traveling back through the canal on an outgoing tide. Though the canal at a moving tide may be hectic, because of the current's strength, the fishing is often good.

Anglers fishing the Metedeconk River's mouth at dawn or dusk often have fun with tailor bluefish. Tailors provide action during the late spring and sporadically throughout the summer while snappers really become a force to reckon with during August.

Did You Know?

Bluefish occur along Africa, parts of Europe, the Mediterranean Sea, Malaysia, and Australia.

Look for snappers to maintain a strong presence within the river and especially at Beaver Dam Creek. The south side of the Metedeconk's mouth, where Herring Island lies east from, is a particularly viable place to catch bluefish.

Anglers should target these waters and also the channel that runs along Herring Island's eastern side. This channel is usually good for anglers who fish early to avoid boat traffic. Anglers should also explore the island's shallows during a high tide.

South of Herring Island, anglers can typically find bluefish marauding baitfish schools at the Mantoloking Bridge and at the mouths of Kettle Creek and Silver Bay. At the bridge, expect to find bluefish near the bridge's pilings while at the creek and bay, look for bluefish to hold near their mouths.

South from Kettle Creek, towards the Route 37 Bridge and the Toms River, anglers should investigate the bay's eastern flank. The water is generally shallow here but drops in some places to 6 or 7-feet deep. Bluefish will often patrol the edges where the water deepens, looking for baitfish, and anglers drifting bunker steaks or working plugs along the edges will get into some action.

While the Route 37 Bridge/Toms River area does not become active with bluefish as soon as locations closer to Barnegat Inlet, one should be able to find tailor bluefish during June and sporadically throughout the summer. Snappers usually provide steady action in this area during July and August. During the summer, snappers will tend to stick close to the bridge, providing light tackle enthusiasts and bobber casters sport. Anglers can find the snappers' larger brethren combing the near-by channel for baitfish. Generally, early morning hours and dusk provide the best action.

Located immediately south from the Route 37 Bridge is the Toms River. Tailor bluefish usually roll into the river during June while late July and August provides great snapper fishing. The Toms River's mouth is a good place to look for blues, with Coates Point, Long Point and Good Luck Point being some of the better spots to try with small plugs and metals. Tailors are not as evident in the river some years, but snappers always seem to maintain a strong presence within the river, especially at the Dillon and Mill Creeks.

South of the Toms River, the bay's eastern side, which becomes Island Beach State Park, should become one's focal point. The bay's western side basically maintains a 7 to 8-foot depth while its eastern waters, save for a few small areas, rises up to form 2-foot deep shallows. These shallows, known as "the flats," basically run along the bay's eastern flank from the Route 37 Bridge south to Oyster Creek Channel. What makes the flats great for bluefishing is they provide protection and food for a wide variety of baitfish, and people for the most part steer clear of the entire area. Thus, bluefish not only have a steady food supply at the flats, they also have an easy route to the inlet and less interference from large, loud objects plowing along the surface.

The two primary spots that attract anglers along the flats are Reeves Cove and Tices Shoal, which are both deeper reaches of water that extend into

The "Flats"

Tices Shoals

Cedar Creek

Holly Park

Stouts Creek

Cedar Creek and Tices Shoals are popular areas to catch bluefish. The shallow water "flats" that exist off Island Beach's western shores hold a very large bluefish population (center). The eel grass flats and marshy areas that exist along the bay's eastern side attract a variety of baitfish and bluefish are never far behind (top left).

the flats. At these two spots, bluefish usually prowl the ledges where the bay's deep water rises to the flats. Savvy anglers simply drift bait or work lures along these ledges for blues throughout the summer. Though Reeves Cove and Tices Shoals receive all the accolades, anywhere along the flats' ledge will typically have bluefish and other gamefish patrolling for prey. Thus, it pays to be investigative! Bluefish will also venture onto the flats, typically during high tide, and anglers who actually venture up onto the flooded flats and locate small depressions will often score well. Just be quiet when approaching a depression because loud

noises from a boat or yourself will often scare the fish.

Across the bay from the flats, anglers can find bluefish around the Intracoastal Waterway, especially off Cedar Creek, Sunrise Beach and Stouts Creek during the late spring and early summer. Around early July, pint-sized snappers begin to terrorize baitfish along the bay's western shores, especially at Cedar Creek and Stouts Creek. Fishing for these rascals really picks up around August.

Lower Bay

With their close proximity to

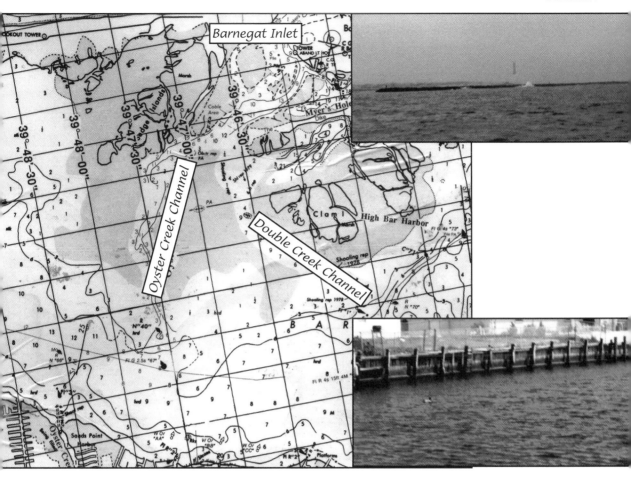

Barnegat Inlet, Oyster Creek Channel and Double Creek Channel are arguably the top places to fish in the lower bay (center). The inlet's north jetty is good for alligator blues (top right) while snappers populate the lower bay's residential lagoons (bottom right).

Barnegat Inlet, from which most bluefish enter the bay, many locations in the lower bay boast solid bluefishing throughout the summer because bluefish have quick access to the ocean. The waters between the Forked River and Island Beach State Park are good to fish during May and June for tailors and larger size bluefish. Though this is a vast area to target, breaking baitfish and dive-bombing birds will often point you in the right direction. Anglers fishing Forked River's mouth score well with tailor blues during the late spring and snappers throughout the summer. Head west into the Forked River for good snapper fishing during the summer.

Many anglers troll small Clark Spoons and tubes off Forked River while on route to Oyster Creek Channel because the water along the western shores maintains a somewhat stable depth of 9-feet.

South of Forked River, one comes upon the Oyster Creek, most noted for the nuclear power plant on its shores. The creek itself provides good fishing for tailor blues and snappers, who journey into the creek after baitfish. Anglers fishing from the Route 9 Bridge score well with snappers and occasionally tailors.

East of the Oyster Creek lies Oyster Creek Channel, which winds easterly through the bay before it meets

Barnegat Inlet. During the late spring and early summer, Oyster Creek Channel provides great bluefish action. As the tide comes in, baitfish begin to filter onto the large eel grass covered flat which Oyster Creek Channel slices through en route to the inlet. This flat then comes alive with bluefish stalking baitfish. During the late spring, anglers can typically stop anywhere along the channel and cast up onto the flat with a Bomber-style plug or a metal and catch bluefish. A few casts should work the flats' interior while a few should be aimed at working the channel's edge. Once the tide begins to recede, bluefish head back to deeper water, either to the channel or to the inlet, to wallop the baitfish being swept off the flat by the receding tide. As the summer wears on, bluefish will still frequent Oyster Creek Channel using it as a route to travel from the inlet to the flats.

The eastern end of Oyster Creek, right before the channel enters the inlet, is a particularly good bluefish spot year 'round because it slopes from the terra firma known as Island Beach down to the 20-feet deep channel in a matter of 5 feet. Couple this with the few shallow creeks, teeming with baitfish, that flow from Island Beach into the deep channel, and one can see why blues stalk this part of the channel. Though weeds tend to choke this area, metals and jigs allow anglers to drop below the weed situation. Oyster Creek also lends itself to chumming, especially at its eastern end. Here, the channel is quite wide, allowing anglers to anchor up near Island Beach and chum into the channel without blocking any boat traffic. Chumming for bluefish at Oyster Creek Channel is usually very effective;

just ask the striped bass anglers diligently dunking their eels and not being able to get past the bluefish.

Rather than follow Oyster Creek Channel into Barnegat Inlet, if you follow it north along Island Beach you will enter the Mud Channel. The Mud Channel is a naturally occurring channel that cuts through the flats behind the Sedge Islands. While this channel is known for its weakfish, it also attracts a sizable bluefish population. The channel runs around 6-feet deep with the light glow of the water to either side reflecting the flats' presence. One problem with this channel is it requires some familiarity with its layout. Though people frequently mark the channel with homemade "unofficial" markers, ranging from stakes to white bleach bottles, these do not usually mark the channel's true route. Though this channel does not receive the boat traffic Oyster Creek or Double Creek receives during an average day, its small size unfortunately permits a single joker plowing through will often scare the fish. In the case of bluefish, they will most likely become dispersed but will remain in the general area whereas weakfish and bass anglers will often have a much tougher time.

If you choose to forgo a trip to the Mud Channel and continue following Oyster Creek Channel east, you will enter Barnegat Inlet. The inlet is probably the best place to catch bluefish throughout the year because it is from there that the ravenous bluefish schools enter the bay typically riding the incoming tide chasing baitfish unmercifully. On the outgoing tide, bluefish simply hang near the inlet's eastern mouth and attack baitfish swept from the bay by the tide. Barnegat Inlet contains many areas that host bluefish, whether they be snappers, tailors or alligators.

Scanning the inlet for likely bluefish spots shows a nice pocket west of the inlet's large sand bar, which is good for snappers and tailors. This sand bar's southern side, where the main channel flows, is a good place to cast lures for roving blues. South from the sand bar lies a trio of bluefish spots: Meyer's Hole, Long Beach Island's marshy banks and the sand flat that brackets Meyers Hole's northern side. These three areas are good to explore early or late in the day.

Heading east towards the ocean, bluefish often frolic right in front of the Barnegat Lighthouse. Anglers can usually find them there either during a moving tide, when the inlet's strong currents buffet baitfish about, or during easterly winds that blow baitfish into the bay.

A stone's throw from the lighthouse, anglers fishing from the south jetty have a good shot at bluefish schools. Land-based anglers should always bring an assortment of metals in case a breaking bluefish school pops up far enough from the angler's vantage point that a light plug has no way of making it. A 1-ounce Crippled Herring, on the other hand, will cast like a bullet and allow anglers to reach fish hanging relatively far away from the south jetty. The jetty's southern side that faces Barnegat Light's beachfront, specifically the dilapidated, "former" south jetty and the "mast," is a good side to fish especially if one fishes from the jetty's tip.

Across from the south jetty, lies the north jetty, complete with jetty jocks casting from its rocks and Winnebagos parked at its base. Throughout the summer and autumn, one can routinely find bluefish terrorizing baitfish schools here. Blues typically hang closer to the north jetty than to the south jetty because the inlet's main channel runs parallel to the north jetty. Thus, the water near the north jetty is deep and swirling while the waters near the south jetty are shallower, evident by their light yellow coloration during low tide. The north jetty commonly has breaking fish its entire length while the south jetty tends to have breaking fish where the water is deeper, either at its tip or where it meets the lighthouse. Lures, whether metals or surface, are commonly used at the jetty with savvy anglers bringing metals because the tides tend to choke the jetty with weeds, making it difficult to employ surface lures, such as Red Fins.

Moving east along the north jetty, good bluefishing usually occurs along the jetty's partially submerged middle. On your initial fishing expedition to the north jetty, it's smart to choose a slack or low tide so you can get a feel for the width and location of the jetty's partially submerged rocks. Ideally, you want to fish the ocean side of these rocks because the inlet's strong currents and boat traffic often makes fishing the inlet side tough even for old salts.

The jetty's ocean side, on the other hand, offers boat anglers a chance to concentrate on fishing rather than navigation, position and the elements. Especially during the autumn, bluefish will often pin schools of bunker or mullet against the north jetty letting

Did You Know?

A 15-pound bluefish is approximately 10 years old.

anglers enjoy solid bluefishing for hours. When migrating bluefish pass by the inlet during the fall, many boat anglers choose to forgo continual repositioning of their boat and simply anchor off the jetty's ocean side near

follow the channel that heads west from the inlet, you will follow Double Creek Channel which offers good bluefishing. The channel's deep stretch located at its eastern end that runs parallel to Long Beach Island, is good to fish though it

these rocks and cast lures, chum or liveline baitfish for visiting bluefish.

The waters off the bell tower, which marks the north jetty's end, are a good but tricky area to fish. The fast current and choppy conditions which usually occur during a moving tide, in addition to the boat traffic, make spot casting and trolling the best choices for this location. Trolling Clark Spoons, Crippled Herrings, diamond jigs, large Hopkins or other metals past the bell tower, especially as you leave the inlet and swing north, is a plausible way to catch a day's worth of blues while on route to your favorite inshore fluke lump or tog wreck.

Heading back to the bay, if you

becomes crowded during the day. Following Double Creek, we see a sand flat guards its northern side for awhile eventually giving way to an eel grass flat, that borders Double Creek for the remainder of its length. Fishing along the edge of the sand and eel grass flats is generally productive for blues because these shallows host a variety of bluefish prey. Thanks to the shallow water, birds will usually pinpoint the bluefish.

Along Double Creek's eastern side, lies Clam Island. Though Clam Island does not seem to produce bluefish as consistently as the flats, one can still do fairly well where Double Creek rises up to form the shallows that surround Clam Island. After Clam Island, Double

Tailor bluefish are very prevalent along the bay's marshy eastern flanks. Investigating this area with surface plug or small metals during dim light conditions will usually result in a fine catch (center).

Creek winds southwest through the bay eventually emptying by marker #68, which provides good bluefishing for lure casters during the late spring. By late summer, snappers are the featured species here.

Leaving Double Creek Channel, the waters from the #68 marker south to Gulf Point is primary ground for late spring bluefish. Though the bottom depth remains fairly constant here, blues still pop up. During the late spring and early summer, look for bluefish prowling the area, giving way to snappers which take up residence throughout the summer. Many of the bottom depressions that attract bluefish also typically hold weakfish. When you find a bluefish school near Gulf Point, try bouncing a bucktail or Fin-S-Fish along the bottom for any nearby weakfish. If snappers are indeed your goal, the numerous residential lagoons carved into the bay's western shores are good areas to check with small baits and lures.

The channel off Gulf Point that heads east towards Long Beach Island's western shores provides bluefishing opportunities throughout the summer. Anglers should investigate the channel's edges and near the few marsh islands that dot this area. Be sure to bring a nice selection of small sized metals if deciding to heave lures for your catch. When fishing this area and Barnegat Bay in general, if you plan your fishing excursions at dawn or dusk, you should be able to enjoy the breaking water bluefish many anglers only witness for a few weeks during the late spring. With bluefish, many times the early bird does indeed get the worm — or at least clearly sees it poking its head from the ground!

Chapter 2
Fluke

Few fish that roam Barnegat Bay are as sought after as the fluke. Also known as summer flounder, Paralichthys dentatus or, in the case of a large fluke, "doormat," fluke are tasty and make a good account of themselves when hooked making them very popular with the angling community. Anglers will find fluke generally receptive to a wide variety of baits while being found in diverse locations.

Distinguishable by their large mouth filled with sharp, prominent, bluefish-like teeth, fluke possess the ability to camouflage themselves against the bottom they are lying against. Thus, what they lack in raw speed they make up for in stealth. Fluke found over a sandy bottom will be a light brown shade while fluke taken over a muddy bottom will be dark brown to almost slate on top with a stark white underside. A trick to know what type of bottom you are fishing over is to notice the color of the fluke you land. Lying camouflaged against the bottom, fluke wait for unsuspecting prey, whether it be a 7-inch snapper or a ½-inch grass shrimp, to glide by and with a quick burst, the fluke is fed.

Though the current IGFA all-tackle record for fluke stands at 22-pounds 7-ounces, the average fluke is around 2 to 5-pounds. Supposedly, fluke are capable of reaching the 35-pound mark; whether a fortunate angler will catch a doormat that big remains to be seen.

Fluke Food

Fluke, similar to striped bass, prey upon a wide variety of marine organisms. The partially digested victims one finds when filleting a fluke, ranging from grass shrimp to baby sea robins, testify to the fluke's diverse diet. One time, my brother found a 2-inch, fully intact hermit crab minus its shell in a fluke's stomach.

Live killies are perhaps the most commonly utilized fluke bait on the Barnegat Bay scene. Whether drifted alone or in combination with a strip bait, killies are a good fluke bait because they are one of Barnegat Bay's predominant baitfish. Killies are especially effective when fished in areas that hold killie populations and are also

Barnegat Bay plays host to many doormat fluke each year. Dick Henderson took this 11-pounder from Double Creek Channel (opposite).

accessible to fluke. One area, for example, is where Oyster Creek Channel whips past the shallow creeks that flow from Island Beach. Killies thrive in the marsh's shallows, and fluke merely position themselves near these creeks and wait either for the tide to flush the killies out of the creek or for the high tide when they can move into shallower water for the killies.

Whether drifted or trolled, killies are best fished alive. Most anglers hook a killie by inserting the hook into the killie's bottom lip and out its top lip. Other anglers hook a killie by inserting the hook into the killie's back or near its tail. This injures the fish, causing it to swim erratically which can entice a wary fluke. Killies are usually fished via either a fish-finder rig, a hi-lo rig or a 3-way swivel rig, with the fish-finder rig being the most popular among bay anglers.

Killies are a very hardy species. Anglers can keep killies alive by either placing them in the standard Flo-Troll bait carrier or on ice and damp seaweed without any water. Whatever you do, don't place killies in a bucket of water for extended periods of time because they will quickly use up the water's oxygen and suffocate.

Just about every area tackle store sells live killies. If you want to catch your own, you can either trap them with the standard galvanized steel trap or net them via a seine net. When considering areas to catch killies, select areas they favor such as the bay's marshes and brackish water areas. Many locations along the bay's western and eastern shores hold killies. Beaver Dam Creek, the Metedeconk River, the Toms River, Cedar Creek, and Oyster

Creek are just a few of many areas where you can catch killies.

For the most part, live bait is essential if one wants to consistently catch fluke or is seeking fluke of doormat proportions. True, strip baits and bucktails in the right hands account for coolers of fluke, but most doormat fluke have let many more strip baits and bucktails go by their head than peanut bunker or snappers. Small 3-inch "peanut" bunker, "finger" mullet, 4-inch snappers, small herring, "silver dollar" butterfish, bay anchovies, and small spot are some baitfish savvy anglers drift and troll each year for fluke. The key to using these live baitfish is to match them to the prevalent baitfish that fluke are feeding upon in your angling location. Fishing live baitfish is especially productive when there is little or no drift. The baitfish's very movements will catch a fluke's eye whereas anglers attempting to drift squid strips get little or no movement from the drift to impart action on the strip. Also, sea robins tend to avoid hitting these larger sized baitfish so anglers can avoid dealing with these characters.

When drifting live baitfish for fluke, anglers tend to favor the fish-finder rig. This rig allows the angler to easily give the fluke line when it engulfs the baitfish so the fluke has time to get the baitfish into its mouth without becoming spooked by feeling any resistance from the angler. Smaller-sized baitfish are typically hooked through the lips killie-style while large baitfish, such as snappers are hooked in the back, slightly ahead of the dorsal fin. Anglers hook these baitfish similar to the way one would hook a herring for striped bass. See page 56 for details concerning baitfish hooking techniques.

Next to killies, squid is probably the most utilized fluke bait on Barnegat Bay. Whether fashioned into strips or rigged whole, squid lends itself to many types of fluke fishing.

Squid strips are a common component to a fluke angler's arsenal whether fished alone or in combination with a baitfish or jig. Fluke seem to favor

inch. The important thing to remember when drifting squid strips, is to make sure the strip flows straight and flutters. You don't want them spinning like a propeller. By simply sliding the strip just beneath the water's surface, you can see whether your strip will drift straight. If spinning, remove the strip from the hook and re-insert your hook. You should hook a squid strip only once, close to its top. Sometimes re-fashioning

native Atlantic squid on many occasions. Tackle stores typically carry this squid, or anglers can jig Atlantic squid off Island Beach with plastic squid jigs. The best strips typically range in size from 3 to 8-inches long and ¼-inch wide. This size strip provides a steady tantalizing flutter that really brings the fluke out of the woodwork. In the right hands, this size strip on a bucktail can really turn heads. To impart more action onto a strip, split the strip's tail end about a ½-

the strip to be sleeker will stop it from spinning.

When using squid strips in areas where crabs abound, savvy fluke anglers spread crab paste or shedder crab oil on their squid strips because fluke in crab-holding locations typically acquire a taste for crab. If in these locations crabs are too numerous, forcing you to change mangled strips every few minutes, you can switch to Uncle Josh pork rind. Crab's will have a

"Keeper" fluke abound in Barnegat Bay's backwater locations, such as Island Beach's western shores. Note the fluke's dark, splotchy coloration which camouflages the fish against a dark bottom (center).

tough time mangling or severing the nearly indestructible pork rind strips. The white bass strip works pretty well, but you should put shedder or shrimp oil on it to mask the rind's unique pork aroma.

Many anglers turn their head from fishing squid in its natural form in favor of cutting it into strips. A nicely cut squid strip is a good fluke bait, but if you want to really target doormat fluke you should try rigging a whole squid instead. A properly rigged whole squid is one of the best baits to tantalize the discerning eye of a doormat fluke while being able to hook smaller mouthed "placemat" fluke. See page 37 to learn how to rig a whole squid for fluke fishing.

Besides squid, anglers fashion strip baits for fluke from a variety of fish. Perhaps the most popular strips are cut from a fluke's white belly or dark back. With current legislation, however, private boaters supposedly can no longer cut strips from a fluke. Apparently, authorities do not want to get into a debate with an angler over whether a 6-inch strip came from a legal or sub-legal sized fluke. Anglers can, however, hack a strip or two from a legal size fluke that is in the cooler as proof if needed. If you have no keeper fluke or want to save all the fluke's meat for a meal, don't worry; anglers achieve good fluke fishing results with strips cut from bluefish, mackerel, mullet, sand shark, sea robin, and spot. Strip baits cut from species that share their habitat with fluke should be used when fishing that area.

Fashion strips from a fish simply by crudely filleting the fish and then slicing the fillet into strips, similar to the technique used with a squid strip. For strips cut from scaly fish such as bluefish or mullet make sure to scale the strip so it flutters. If the strip is spinning

Fashioning Strip Baits from Squid

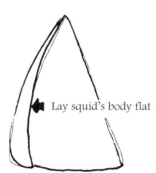

Lay squid's body flat

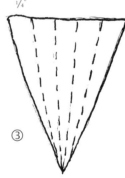

¼"

① To fashion strips from a squid, simply slice the squid's tubular body length wise along its center, being sure not to sever the squid in half. Remove the squid's head and wings.

② Take either side of where you just cut and spread out these flaps so that you have one large flat piece. Remove the various organs and head so that you have a flat V-shaped piece.

③ Starting at the "V's" flat top, start your cut ¼-inch from the edge and slice on a downward angle aiming towards the "V's" point. Fashion strip baits from sea robins, mackerel and other fish fillets this way also.

rather than running straight, thin the strips out by removing some of the meat attached to the skin. Particularly with mackerel, experiment with whether to drift the strip's bright silvery skin side facing the bottom or the darker meat side facing the bottom. Fluke have their preference but that preference often changes according to water clarity and time of day. Good strips for bay use are in the 4 to 7-inch range. When fishing longer strips in the 9 or 10-inch range, anglers avoid missing smaller but legal size fluke, that usually bite the strip's rear by using either a tandem rig, two hooks tied in tandem with one hook inserted at the strip's beginning while the second or "stinger" hook takes up the rear accounting for the short strikes, or a ryder style hook, a long shanked hook with two hooks forged to the same shank, one hook is inserted at the strip's top and the second hook at the strip's end. Similar to squid strips, strip baits are best employed when drifted or trolled. If you find a driftless day or adverse conditions, either troll your strips or abandon them in favor of either a bucktail or a lively snapper.

Some baitfish fluke relish eating are delicate and thus hard to keep alive. Spearing and sand eels are two baitfish more commonly drifted or trolled dead. By using baitfish in fresh condition and avoiding frozen or rancid baitfish, anglers can have a productive outing.

Hooking dead baitfish for drifting or trolling is best done by inserting the hook through the fish's lips. When fishing long baitfish particularly sand eels or large spearing, you may have to straighten the fish from rigormortis by breaking its backbone.

Similar to a strip bait, you want the baitfish to swim straight and not spin. When fishing dead baitfish alone or in combination with a strip bait, be sure to bounce your sinker along the bottom to impart movement on the baitfish.

Grass shrimp are perhaps the most abundant fluke prey found in Barnegat Bay. When filleting fluke, remnants of a grass shrimp dinner are often found within a fluke's stomach. A good way to use a fluke's love for grass shrimp to your advantage is to accent your strip baits or bucktails with shrimp oil. The bottled tackle store variety works well but if you have the time, it is better to employ fresh shrimp oil. Net around a salt marsh's banks, eel grass flats, bulkheadings, and pilings with a fine mesh scoop net to catch yourself a small bucket of grass shrimp. Crush the shrimp and marinate your bait or bucktail in the crushed shrimp bodies. Anglers employing strip baits or bucktails saturated with shrimp oil near areas that host both shrimp and fluke typically score well with fluke while putting a dent in the local weakfish population.

Tackle and Techniques

The rig most commonly utilized by Barnegat Bay fluke anglers is the fish-finder rig. This rig is ideal for fluke fishing because it allows the angler to feel the fluke engulf the bait without the fluke feeling resistance from the angler. When a fluke hits the bait, the angler points their rod tip at the fluke while free spooling line it. The fluke, meanwhile, feels nothing because the

Did You Know?
An immature fluke possesses an eye on either side of its head like a bluefish. But, as the fluke matures, one eye gradually moves to lie beside the other eye.

rig's sinker stays stationary while the free spooled line flows unhindered through the fish-finder. After two or three seconds, the angler stops free spooling line, waits for the line to go tight and sets the hook with a quick snap of the wrist followed through by a fast vertical thrust. The fish-finder is an easy rig to tie and can be quickly altered if conditions change.

Another popular rig that anglers drift for fluke with is the hi-lo rig. This rig lets the angler work one bait along the bottom while simultaneously testing the water a few feet higher off the bottom. Anglers typically fashion this rig so the "hi" bait floats about 3-feet off the bottom for weakfish or bluefish while the "lo" bait drifts along the bottom for fluke.

Anglers targeting doormat fluke frequently employ large baits during their quest. The rig for drifting or trolling live baitfish, such as mullet or snappers, is relatively simple. It is basically a fish-finder rig with an appropriate size wide gap hook, such as a #4, or a treble hook. Most baitfish are hooked either through the lips killie-style or slightly before the dorsal fin, inserting the hook only $1/8$-inch into its back to allow maximum movement

Popular Fluke Rigs

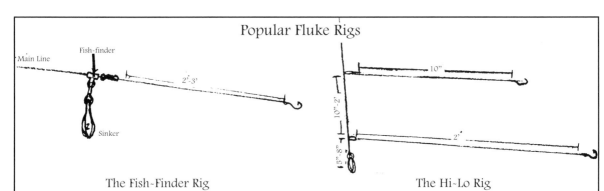

The Fish-Finder Rig

To fashion the fish-finder rig, snell a English wide gap hook to one end of the leader. Hooks range in size from 3/0 used for large fluke to a 6/0 when smaller fluke are targeted. Fluke have the tendency to engulf baits so a smaller hook is generally not the answer because most of these hooks will wind up in the fluke's gullet. Leader lengths range in size from 3-feet, which are used in areas where there is a swift current or during windy, fast drift conditions, to 2-feet, which is used on both slow or fast drift conditions.

You then want to take a Luxor fish finder and run the line from your reel through the fish finder's white plastic sleeve. With the fish finder riding beyond the end of your line, attach a swivel to the line. The swivel will prevent the fish-finder from sliding down the leader. Now, attach the leader to the swivel.

Once completed, the fish finder, with a bank sinker attached for bay fishing will slide to the barrel swivel and stop against it. There will then be no hardware save the swivel separating you from whatever takes your bait. Sinkers range in size from $1/2$-ounce to 5-ounces depending upon the location and the conditions.

The Hi-Lo Rig

For the hi-lo rig, take a 36 to 48-inch piece of 15 to 20-pound leader material and tie a swivel at one end and a surgeon's end loop at the other end. About 5 or 8-inches above the end loop, tie a drooper loop. Tie a second drooper loop 10 to 24-inches higher than the first loop. You then snell two English wide gap hooks, one to a 24-inch piece of leader and the other to a 10-inch piece of leader. Tie the longer leader to the bottom drooper loop while tying the shorter leader to the top loop. The leaders' different lengths will prevent tangles between the hi hook and lo hook.

Bait the "lo" hook with a fluke oriented bait, such as a live killie, while baiting the "hi" hook with a bait aimed at weakfish or bluefish, such as a squid or mackerel strip.

while giving the angler a sure way to firmly hook the fluke when it hits the baitfish. Though rigging a whole squid for fluke fishing is a tad more complicated than simply inserting a hook into a baitfish, a whole squid will surprise you at how effective they can be for doormats, especially during the fall.

To rig a squid for fluke fishing you will need: a squid with an overall length preferably in the 6 to 8 inch range, a rigging needle, 20-pound mono leader material, a 12-inch piece of 20-pound dacron line, a ryder style hook, a #10 English wide gap hook, and a little patience!

Fresh "Jersey squid" works best for rigging due to the firm consistency of its flesh. Small "coaster" fluke do not tear Jersey squid apart as easily as they do the boxed tackle store variety. So, it is possible to salvage a few baits after encounters with the "coaster" fluke that will aggressively attack a rigged squid that is sometimes half their length. The ideal time to rig a squid, without mangling its flesh and sprouting a few gray hairs, is to rig it while it's not fully defrosted. The squid should be hard enough so it resists loosing all of its fluids but soft enough so the rigging needle can penetrate it without much pressure.

Construct the initial part of the rig by snelling the ryder hook to a 36-inch piece of the 20-pound mono leader material. Make sure the leader material is free of any abrasions that could come back to haunt you while fighting a doormat. Tie an surgeon's end loop on the other end of the line. Set this part of the rig aside.

A fluke's point of attack will usually center around the squid's head area so include a "stinger" hook with your rig to augment the probability of hooking a doormat and especially the feisty but relatively small mouthed placemats. For the stinger hook, snell the #10 wide gap hook to the 12-inch piece of 20-pound dacron line. Mono leader material tends to be a little too stiff for rigging purposes while not offering dacron's solid no-stretch capabilities. Take the other end of the dacron line and thread it on the rigging needle. In lieu of a rigging needle, employ any long, thin somewhat sharp object such as a piece of aluminum wire. Insert the needle in the squid's mouth cavity and run it up through the squid's head being careful not to deflate one of its eyes. Push the needle out the squid's mantle about an inch away from where the bottom hook of the ryder hook will be set. Disconnect the rigging needle from the dacron line.

Pull the dacron line so the stinger hook is hidden amongst the tentacles. Ideally, you want to set the stinger hook so that the hook's eye rests right in the squid's beak. This carefully tailored placement offers a nice solid connection to the main rig plus an added camouflage quality that will trick a doormat that has eluded capture by the angling community.

Rig the squid to the Ryder hook rig, previously set aside, by first centering the Ryder hook on the squid's mantle. Insert the Ryder's first hook at the top part of the squid's mantle where it forms a point making sure that the hook is centered properly. Both of the Ryder's hooks should not penetrate completely through the squid's body but rather pierce the skin and remain within the body cavity. The Ryder hook's proper placement will cause the squid to "swim" straight during drifting. Connect the stinger line to the Ryder hook via a uni-knot. This knot is relatively easy to tie while possessing superior holding qualities.

The Ryder hook should be firmly against the squid's mantle somewhat camouflaged against the squid's skin while the stinger hook, hidden in the mass of tentacles, is deceived only by an occasional glint. During drifting, if you rigged the squid properly it should "swim" straight with its tentacles offering an occasional flutter.

When selecting a rod to drift for fluke, you want the rod to basically be an extension of your arm. It should possess a stiff tip so you can easily set the hook but sensitive enough so you can quickly detect a fluke. Stiff-tipped 100% graphite rods between $5\frac{1}{2}$ to 7-feet rated for 10 to 25-pound test are good choices for the bay. Shimano, Fenwick and Berkley are a few distributors whose rods fit the bill. Many anglers fit these type rods with small freshwater style conventional reels, so they can easily free spool line as soon as they feel a fluke. Most spool their reels with line between 8 to 12-pound test for bay fluke fishing.

When fluke take a drifted or trolled bait, they usually feel like a dead weight. You will be bouncing your bait along the bottom, and then it will feel like you snagged the bottom or a large crab grabbed your bait. When you feel anything remotely like this, free spool about 5 to 10-feet of line to the fluke so the fluke can get the bait fully into its mouth. Engage your reel, point your rod tip at the fish, extend your arm, and

wait for the line to become tight before you drop the hammer. For smaller baits, such as small killies or squid strips, either drop your rod tip or ladle out only 3 to 5-feet of line before striking back. When fishing larger baits, such as long, meaty mackerel strips or live snappers, stick to giving 10-feet of line so the fluke can get the baitfish fully in its mouth before you set the hook. In both scenarios, set the hook with a quick snap of the wrist followed through with a sharp vertical lift. Make sure your reel's drag is set properly and have fun.

Trolling

All fluke anglers have experienced days when they are excited to drift for fluke only to find either too little or too much wind preventing any worthwhile drifting in their favorite spot. Other times, the wind is fine, but it blows in the opposite direction of the current which basically stabilizes the boat. Days like these literally separate the novices from the experts. Novices either pack it in or try a half-baked attempt at drifting while experts merely decide to burn a tad more gas than they originally planned and troll the area for fluke. Even on days that are ideal for drifting, savvy anglers drift to locate fluke and then employ trolling to continually position their baits along the small amount of bottom where the fluke are holding. Meanwhile, drifters largely spend their time drifting over unproductive bottom and repositioning their boats.

When trolling or "power drifting," you are basically creating your own drift via your engine rather than relying on the wind or current for movement. Trolling for fluke is rather simple, all you do is drive the boat very s-l-o-w-l-y over typical fluke holding areas, such as channel edges, the down-tide slope of flats and creek mouths, so your bait will "drift" very slowly along the bottom.

On stagnant days, you can work a location from many different angles because no natural force will deter you from your course. When faced with adverse conditions, you want to face your boat into the current or wind and make the boat go as fast as it takes to move the baits slowly along the bottom. You are basically creating your own drift so figure out the best speed. Also, keep the boat going somewhat straight so multiple rods do not become tangled.

The standard fish-finder rig coupled with the conventional rod, described earlier, is standard for trolling. It allows the angler to easily free spool line to the fluke without the fluke feeling the boat's movement. Some anglers forgo a fish-finder rig and employ a 3-way swivel rig. This rig basically consists of your main line tied to the swivel's top eyelet, a 3-foot leader with snelled hook tied to a bottom eyelet while a bank sinker tied to a 3 to 5-inch piece of 10-pound test is tied to the third eyelet. Depending on your trolling speed, you may opt to shorten the leader of your rig from the standard 3-foot length to 2-feet or even 18-inches to get your bait closer to the bottom.

A Buck Tale

Heading the short list of artificial lures employed for fluke in the bay is the bucktail. A bucktail is a leadhead jig tied with hair to conceal the jig's hook. In the past, the white hair from a deer's white tail graced bucktails, hence the name "bucktail." Anglers bounce bucktails slowly along the bottom of likely fluke holding areas, such as the slopes of the flats that flank Oyster Creek and Double Creek Channels, the drop-offs of the deep holes off the Mantoloking Bridge, the pilings of this bridge or the Route 37 Bridge, and the creek mouths that dot Island Beach. All are prime locations for jigging bucktails.

Bucktails are differentiated by the shape of the jigs' leadheads. Common bucktails are: lima bean, open mouthed smilin' bill, bullet shape, torpedo, bug eye and cannonball. Bucktails come in a rainbow of colors, with pure white being the most popular selection.

Within the bay, anglers use bucktails from $\frac{1}{8}$th of an ounce to 3-ounces in weight. One-eighth to $\frac{1}{2}$-ounce bucktails are used in areas where small baitfish, such as spearing, abound. The larger 2 and 3-ounce sizes are used either at locations with strong currents, such as the Point Pleasant Canal's mouth or the inlet, or as a sinker substitute for a fluke rig. Whatever weight bucktail you choose to employ, avoid using wire leaders or swivels. These items hinder a bucktail's action. Simply tie the bucktail directly to your line for the best results.

Bucktails are rarely if ever used without a "sweetener." A sweetener is an item, frequently a strip bait, that an angler attaches to a bucktail to add movement and scent to the jig. Freshwater jelly worms, plastic grubs, strip baits, live killies, sea worms, and shedder crab have all been used as a

Did You Know?
Because the bucktail is a potent lure for a wide variety of fish, it was chosen to be included in every military aviator's survival kit during World War II.

When fishing for fluke, remember to bring needlenose pliers or some other instrument to disgorge a hook from a fluke's tooth filled mouth.

bucktail sweetener at times. Among the bucktail sweeteners chiefly aimed at fluke are: squid strips, fluke belly, mackerel strips, sea robin belly strips, shedder crab, large sand eels and Uncle Josh pork rind. My personal favorite bucktail sweetener is a squid strip about $\frac{1}{4}$ wide and 3 to 4-inches long.

The best rod for jigging bucktails are stiff-tipped graphite rods $5\frac{1}{2}$ to $6\frac{1}{2}$-feet long rated for 6 to 12-pound test equipped with spinning reels. Reels are typically spooled with 6 to 10-pound test. The rod's stiff tip is essential for bucktailing because it allows you to set the hook immediately when you feel the fluke before the fluke spits the bucktail out. When bucktailing, you want to try to use light line, perhaps 8-pound test, because lighter line has less drag in the water which lets you jig the smaller, lighter bucktails in fast moving locations. If the fluke are feeding on small baitfish, a large, heavy bucktail won't cut it while small bucktails, which can be jigged along the bottom with light line, will score. While bucktailing at areas with strong currents, such as the inlet, you may have to experiment with different sizes before you arrive at a weight that will keep you on the bottom.

To work a bucktail, you generally want to cast the bucktail up tide from your position and free spool line until the bucktail hits bottom. You then engage your reel, reel in your slack line and jig the bucktail slowly by raising your rod tip about 6-inches and then lowering it. When lowering your rod tip, you should feel the bucktail hit bottom. Jig the bucktail slowly so it pokes along the bottom in little 1 to 2-inch hops, similar to how a spearing

would poke along the bottom near a fluke's mouth. You always want the bucktail to keep close contact with the bottom. Working the bucktail quickly so it leaps like Superman from the '50s is not typically too productive for fluke — bluefish, maybe! Fluke will usually pounce on the bucktail when the jig is fluttering back towards the bottom. The fluke's hit will usually telegraph up your line as a dead weight or as an aggressive, strong pull. Once you feel a hit, it is pivotal that you quickly set the hook with a quick but strong yank before the fluke spits out the bucktail. A fluke is not likely to sit on a bucktail as long as it would a live baitfish.

Besides Bucktails

Besides bucktails, there are a few other lures that are very effective at catching fluke. Popular lures employed for fluke include: Fin-S-Fish, jelly worms, shad darts, Berkley Power Sandworm, and DOA shrimp.

Known more amongst weakfish anglers, the Fin-S-Fish has recently become a popular lure for fluke. A Fin-S-Fish is a jelly fish, that comes in an array of colors and sizes, and is threaded onto a small leadhead. Anglers typically jig these lures bucktail-style for best results. Best colors for Fin-S-Fish are ones that mimic the coloration of the forage fluke feed upon. Colors such as alewife shad, gold pepper shiner and chartreuse ice usually work well, but it pays to experiment because a location's prevalent baitfish and water clarity will determine what colors will succeed.

Mann's jelly worms, Berkley Power Sandworms and plastic twister tail grubs placed on a plain leadhead and jigged along the bottom similar to a bucktail work well for fluke, though not usually as productive as a bucktail.

Small plastic shrimp, particularly the DOA shrimp, work fine for fluke if you alter the standard bucktail jigging technique to resemble a shrimp's natural movements. Mix in a few short, swift jigs with the slow bounce to mimic a shrimp's quick but short movements. Many times you will catch more weakfish than fluke with these shrimp.

Where and When

Fluke occur from Maine to South Carolina with their greatest concentration being found between Cape Cod, Massachusetts and Cape Hatteras, North Carolina. Lucky for us New Jerseyites, we live right in the middle of this concentration.

Fluke are bottom dwellers. They can be found lying on mud, sand, a mud/sand mix, or gravel bottoms. If there is one consistency to fluke, it is that they can always be found in locations that attract and hold baitfish. Fluke live by ambush not by superior open water speed so they will naturally position themselves in areas where they can lay camouflaged against the bottom and engulf their unwary prey. Such areas typically take the form of: a channel edge, a creek or river mouth, the edge of a sand bar, an eel grass flat, a deep hole, the outskirts of a rock pile, and a shipwreck site. Many fluke will position themselves downtide of such areas because when the tide recedes, it brings baitfish off the flat to the deep, often turbulent waters where the

baitfish's speed and maneuverability are significantly hampered. Once a shallow water location floods during an incoming tide, fluke will actually venture into these shallow water areas to go after baitfish. They then retreat when the tide begins to recede to wait for baitfish to get dragged from the shallows.

Besides baitfish activity, water temperature will determine where fluke position themselves in the bay. Fluke seem to prefer water temperatures in the mid 60s to low 70s, being sensitive to the bottom temperature not the surface temperature. There won't be a significant difference in surface and bottom temperature at a shallow location such as Tices Shoals, however a 20-foot deep hole in the inlet is a different story.

Besides depth, the tides play a major hand in determining a spot's water temperature. In back bay locations, such as Gulf Point, the tide's effect on the water temperature is not as obvious as it is at the inlet. If you ever swim in the inlet during an outgoing tide, you will find the water usually warm and murky because it's from the bay. During an incoming tide, the water comes directly from the ocean and is noticeably cooler and cleaner. A sharp fluctuation in water temperature has a pivotal effect on fluke. If the incoming water cools an area down too much, fluke will not feed as strongly as if the water's temperature was more to their liking.

Remember fluke are a schooling fish, just like bluefish and winter flounder. This means that once you catch one, you should continue to fish the area because where there is one fluke, there are usually many more.

Barnegat Bay's fluke season usually kicks off around early May as the first fluke enter the bay through the inlet. As May wears on, fluke begin to fill the bay with their greater concentrations centering around Barnegat Inlet and Point Pleasant Canal. Fluke fishing in the bay remains good throughout the summer with late June and early July usually offering some particularly good fishing throughout the bay. The fishing tends to pick up during late August and early September as fluke begin to leave their summer homes. During September, anglers concentrate their efforts at the inlet to head-off the fluke exiting the bay for their oceanic winter grounds. Anglers drifting the inshore waters off Barnegat Light, at areas such as Harvey Cedars Lump, will catch fluke some years well into October.

Upper Bay

Perhaps the best location to fish for fluke when considering the upper bay is the mouth of the Point Pleasant Canal. The canal's mouth is a prime area for fluke because the mouth is where the tide brings baitfish into the bay from the Manasquan River. The holes and slightest bottom contours that occur south of the canal's mouth hold fluke, which will simply lay and feed on the baitfish pushed by the tide. One particular area to drift or troll is a deep hole located a few hundred feet south of the canal. Anglers should stick to working the actual hole rather than its slopes. While at the canal, anglers should also try drifting along the channel that flows from the canal, especially along the channel's edges.

Before bidding adieu to the canal mouth area, anglers should explore the waters directly in front of the Bay Head Yacht Club. Here, the channel drops below the 15-foot mark, and fluke typically hang along the channel's edges or dead center in the channel's deep pockets waiting for the tide to sweep baitfish by.

Southwest of the Point Pleasant Canal flows the Metedeconk River which hosts a diversity of fluke prey, and not surprisingly fluke. Anglers can find healthy amounts of fluke at the river especially from mid-June to late August. Fluke tend to gravitate at the river's mouth, specifically downtide or within any small pockets, bottom contours and channel edges that occur. As one travels upriver, small "coaster size" fluke become more of the norm. One area of the Metedeconk worth investigating is the water between the Metedeconk's southern tip and Herring Island's western shores. Fishing the uneven bottom here during a moving tide is usually productive. You will find that many of the areas in the Metedeconk that hold winter flounder during the winter attract fluke during the summer. One such location, that doubles as a winter flounder and fluke habitat, lies just east of the Metedeconk — Herring Island.

Besides fishing the small channel that squeezes between the island's western side and the Metedeconk's southern tip, anglers should try the island's northwest tip. This tip, which lies directly east of the Metedeconk's mouth, is generally productive when fished with a little patience during a high tide so anglers can work previously shallow water.

Not too far from here, most anglers prefer to fish the island's eastern side where the channel from the Point Pleasant Canal flows near. Anglers drifting or trolling slowly along the channel's edges generally score well with fluke. One particularly attractive spot is where Herring Island juts out to the east. Here, the channel drops past

Good fluke fishing can also be had at the few deep holes that lie south of the bridge. These holes, favored by winter flounder anglers, are deep and within close proximity to the bridge and the forage it attracts. While the surrounding water near these holes is around 6-feet deep, a few of the holes drop to 20-feet deep. Anglers who drift

the 10-foot mark, and the "early birds" work the deep water. A tad south from Herring Island stands the Mantoloking Bridge.

The Mantoloking Bridge is a good location for fluke because the bridge's algae encrusted structure offers fluke forage "apparent" security and food. This forage, in turn, provides fluke with a steady food supply. When fishing the bridge for fluke, anglers typically fish the channel that flows under the bridge, focusing on the channel's edges and especially the bottom near the bridge where baitfish saunter and fluke set up ambushes.

or troll the edges or deepest reaches of these holes generally catch fish, provided that a previous angler did not clean the hole out. If drifting in a southernly direction from the holes, try a drift east of Swan Point before backtracking to the bridge or going elsewhere.

The next notable fluke area after the Mantoloking Bridge is the Toms River area, which encompasses the Route 37 Bridge, Pelican Island and the Toms River. Drifting the channel that winds south beneath the Route 37 Bridge and past the Toms River is a good tactic. East of the bridge, many anglers

"It's no fluke" that anglers who fish the slightest bottom contours catch fluke. The author took this 20-inch specimen over "seemingly" flat bottom near Double Creek Channel (center).

work a thin channel, 6 to 7-feet deep, which runs parallel with the Route 37 Bridge and extends east towards Island Beach. This channel's edges are good for drifting, trolling and jig bouncing because the channel slopes from 7-feet up to about 2-feet, and fluke hang right along the edge. Trolling this channel east will eventually lead you to Pelican Island.

The waters that lie on either side of Pelican Island are good areas to drift for fluke. Many anglers fish off the island's northwestern tip because a depression located here often holds a sizable fluke population. One area in particular that fluke anglers should consider is Pelican Island's southern side. Here, there is a small channel that snakes around the island. Before it curves south and rises to about 6-feet, the channel contains a few fluke-holding pockets that drop past 10-feet deep.

Southwest of Pelican Island flows the Toms River. Similar to the Metedeconk River, most of the Toms River's fluke tend to stay near its mouth. The waters between Coates Point and Goodluck Point are generally productive during the summer. One particular area anglers should investigate is the water northeast from Goodluck Point. Here, the water is 6 to 7-feet in depth with shallows bordering its southwest side, Goodluck Point bordering its northeastern side, and the Toms River lying south. When the outgoing tide and the river's natural flow pushes baitfish east, fluke usually hang in this relatively deep water to ambush them.

Traveling west into the Toms River, productive fluke fishing can be had as far west as the Island Heights area. The spots off of Long Point, which

are targeted by winter flounder anglers, and at the mouth of the Dillon Creek can be good places to check out. When fishing the Toms River, plan on fishing very early in the morning to avoid the river's often hectic boat traffic. Also, keep your eye on any strong rains. One good three day monsoon can dirty the river to the point where it's better to fish closer to Barnegat Inlet where the water is usually cleaner.

The next big fluke area south of the Toms River is Cedar Creek. Known more for its weakfishing than fluke fishing opportunities, Cedar Creek, which lies northwest of Tices Shoals is a good place for fluke. Anglers should stick to working the creek's mouth where the water depth drops from the 2 to 3-foot "creek bottom" down to the 6-foot "bay bottom." Fluke along with a host of other predatory fish typically hold here during an outgoing tide for baitfish, shrimp and crabs to get flushed out by the tide. The #40 buoy off Cedar Creek is another popular area for fluke anglers. Smart anglers start early in the morning when fishing Cedar Creek to avoid heavy daytime boat traffic, especially during the weekend.

Across the bay from Cedar Creek lies an unsuspecting fluke spot — Tices Shoals. This mecca for weakfish also attracts fluke near the same shoal ledges weakfish patrol for baitfish and shrimp. Drifting squid on a hi-lo rig or employing bucktails gives fluke anglers the opportunity to score with weakfish as well as fluke. Be forewarned: Tices Shoals often becomes crowded with boats and grass. Insects can also become a nuisance, especially on stagnant days.

Lower Bay

The first two notable fluke angling locations in the lower bay are the Forked River and Oyster Creek. Anglers drifting the Forked River's mouth generally put together good catches. They typically start their drift off the river and then drift either south to Oyster Creek Channel or north to Stouts Creek, depending on the tide and wind.

Just south of Forked River lies Oyster Creek. Fluke will often enter the creek, and anglers fishing from the creek's mouth west to the Route 9 Bridge will generally do quite well. Though the fluke become smaller as one heads west into Oyster Creek, this should not deter anglers from any explorations. For the most part, the better fluke fishing in the Oyster Creek area is to be had at Oyster Creek Channel, which lies east from the creek.

Oyster Creek Channel is a prime area to find fluke because it is a relatively deep stretch of water dredged for boat traffic that cuts through a shallow, eel grass flat and flows past Island Beach while on route to Barnegat Inlet. Buoy #67, that marks the channel's western mouth, is a popular area to drift for fluke. This area's bottom contains many slight contours that go unnoticed by many anglers but not by too many fluke. Though you may only catch one or two keeper sized fluke per contour, find four contours, which is not exactly a herculean effort, and you will have your bag limit while fishing in a relatively confined area. One thing though, fish this area either very early in the morning or during the week because this place will turn into a zoo on a sunny weekend day.

Heading east into Oyster Creek Channel, fishing the edges of the channel as it snakes towards the inlet or drifting north to south across sections of the channel are good ways to catch fluke. Obviously, some areas of the channel produce more fluke than others, but when your honey hole does not produce or is unfishable due to boat traffic or wind, merely move east or west along the channel and fish another section. With Oyster Creek Channel running roughly 10-feet deep while being bracketed by 3-foot deep eel grass flats to the north and south, you should try drifting or trolling along the channel's edges. Though you are generally best off fishing the channel's edges, the channel's middle part often produces quite well if you are able to drift over any bottom contours.

Rather than following Oyster Creek Channel into Barnegat Inlet, if you initially follow Oyster Creek and continue north along Island Beach, you will enter the Mud Channel. This relatively small "hidden" channel runs along Island Beach's Sedge Island shores and holds fluke in addition to other fish species. Fishing the Mud Channel with a large boat is tricky because it is not that wide nor extremely deep. If you own a boat with a low draft, such as a row boat, then you can really take advantage of the channel's good fluke fishing. If you follow the formula of working the channel's edges and deep depressions with local forage, particularly bay anchovies, spearing or sand eels, you will score here. For bay anchovies, which are prevalent and usually quite small at about $3\frac{1}{2}$-inches, try using a small, light hook such as a #9 or #10

Did You Know?
New Jersey's state record for fluke stands at 19-pounds, 12-ounces. This fish was taken off Cape May and is one of the state's oldest records being set back in 1953.

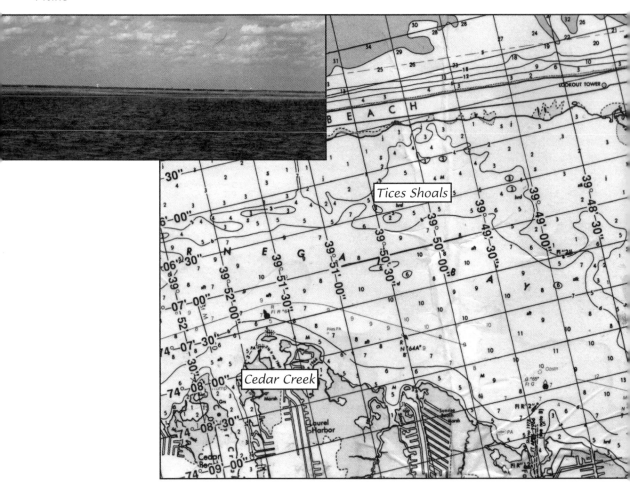

Tices Shoals

Cedar Creek

Cedar Creek and Tices Shoals are good areas to locate fluke when considering areas north of Oyster Creek (center). The salt marshes that line Island Beach's western shores are prime locations for fluke, especially if one can locate a creek flowing from the marsh into the bay (top left).

wide gap so the anchovy is not weighed down by a heavy hook. Savvy anglers can net fresh bait, which almost always equates to the area's predominant baitfish, in Island Beach's creeks or the shores of the nearby Sedge Islands. Baitfish that would normally be a bear to keep alive can be quickly used after being caught.

Oyster Creek Channel's eastern extent, that runs parallel to Island Beach before it merges into the inlet, is a very popular area to fish for fluke. Here, Island Beach's shallow creeks that teem with baitfish empty into Oyster Creek Channel. Anglers typically score well

with fluke right along the ledges where the shallow creeks empty into the 15-foot deep channel. Also at the channel's eastern mouth, anglers fishing the gradual slope of the sand flat that lies across from Island Beach typically do well with fluke.

Oyster Creek Channel eventually merges with Barnegat Inlet. Barnegat Inlet is probably the most popular fluke fishing location in Barnegat Bay. At the inlet, anglers get first crack at the bay's fluke during early May and last licks on the fluke sometimes as late as early October. Late May through early September are

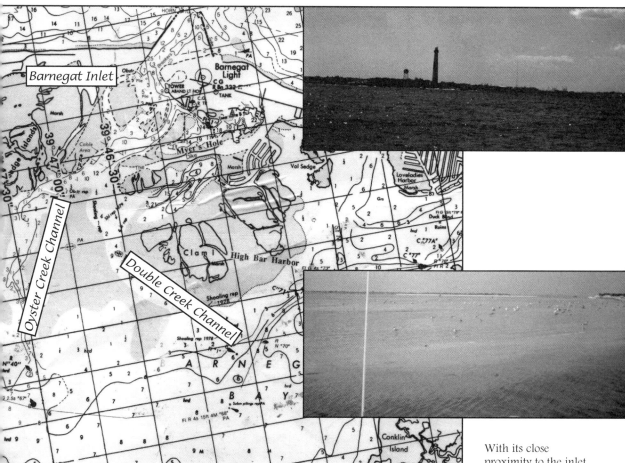

With its close proximity to the inlet from where the bulk of the fluke enter the bay, the lower bay encompasses arguably the bay's better fluke fishing areas (center).

The inlet's main channel is one of a few spots that provide good fluke fishing (top right). Seeing baitfish scatter across the water's surface near a sand bar should make you think that somewhere near the sand bar are fluke camouflaged against the bottom waiting (bottom right).

generally the better times to fish the inlet.

The inlet's western mouth, where the inlet meets the bay is a good area to start a trip. When fishing here, bring heavier sinkers and bucktails than you would normally employ in the bay because the current is strong during a moving tide. When fishing the inlet, bring a few 4 and 5-ounce sinkers to be on the safe side if the tide begins to run and you want to drift.

Though you can get into trouble drifting dead center in the inlet's main channel, making quick drifts across it early in the morning can be productive. The part of the channel that lies directly in front of the lighthouse typically attracts anglers because the water here is deep and the bottom contains much variation.

If you follow the main channel south towards Long Beach Island, rather than north towards the ocean, the channel will deepen as you near a small marsh island. This channel, which Barnegat Light's party boat fleet uses to reach the ocean is good for fluke, especially close to the island. If you head northwest from the island, you'll enter Meyers Hole which holds fluke in addition to weakfish and blowfish.

Northeast from Meyers Hole, early birds frequently target the inlet's main channel where it flows past the north jetty before it enters the Atlantic. If you arrive early enough, drifting just off the jetty is productive when there are no sunami-like wakes to wash you and your boat onto the rocks. Fluke often hang close to the north jetty but within the channel and pick off the baitfish buffeted by the inlet's strong current. Once the boat traffic or wind picks up, productive fishing tapers off unless you are after sea robins. When drifting off the north jetty, keep a casting rod handy for the bluefish that frequently pop up near this jetty.

Across from the north jetty, one will find the yellowish-shaded waters around the south jetty much shallower. The trick to fluke fishing the south jetty is to locate small pockets and contours where fluke will lie downtide from waiting for baitfish. When the tide is moving at its strongest, it is usually better to head back to the bay or venture into the inshore waters off Barnegat Light.

For boaters, the immediate inshore area off Island Beach and Long Beach Island offers excellent fluke fishing. Anglers should investigate the waters off Island Beach's Coast Guard station, the governor's mansion and Long Beach Island's Harvey Cedars Lump. Anglers might also want to explore the inshore wrecks that lie off Island Beach and Long Beach Island. See page 99 for a list of a few wreck sites close to Barnegat Light that usually hold fluke.

Back to the bay, anglers should investigate Double Creek Channel, which exits Barnegat Inlet from the west. Double Creek Channel, which winds through the southern-most portion of Barnegat Bay before it enters Barnegat Inlet, has numerous areas that attract various types of fluke forage making it one of the top areas to find fluke when considering the Barnegat Bay area.

If you enter Double Creek's eastern end and head south away from Barnegat Inlet, you will notice the channel is bracketed by a salt marsh to the left and a sand flat to the right. Although buoys mark Double Creek's basic route and width, they do not account for the true course of the channel in some parts. In particular, the buoys do not account for a deep stretch of water that runs parallel to the front of the salt marsh. Though this stretch may be officially out of the channel, parts of it exceed the 15-foot mark a scant 3-feet from the shoreline.

Across from the salt marsh, the water's soft gold tint points out the sand flat that borders Double Creek's western side. The channel's sharp upward slope to the sand flat is another favorite fluke haunt but be careful not to drift aground because this area has the tendency to slope from 9-feet up to terra firma in a matter of yards.

As you continue south, Double Creek deepens at the junction point where a small residential channel extends south along Long Beach Island while Double Creek veers off to the west. Considering that this junction area dips past the 15-foot mark while the bottom rises to about 6-feet a few yards away, it's easy to see why anglers have a field day with the fluke at this location if conditions are right. The boat traffic and narrow width that characterize the junction point makes drifting a little tricky at times. This area also tends to get choked with weeds, as Double Creek does on a whole, after a storm and usually during an outgoing tide.

As you turn west and head through the junction, the sand flat to the north presents a sharp contrast to the dark churning water that flows past it. The flats on either side of Double Creek partially shield the channel from strong winds so it's possible to drift parallel to the sand flat during a moving tide in otherwise belligerent weather conditions.

After a few hundred yards Double Creek veers to the south, keeping a steady bottom with an occasional dip in depth for the remainder of its length. The shallow eel grass flats that flank Double Creek in this stretch warrant investigation because grass shrimp and other fluke forage infest these shallows and hardly any boats fish these locations. Fluke do not reach doormat proportions by laying exactly where it is convenient for us to fish! Double Creek eventually ends by the #68 marker. Though this area tends to attract more crabs and weakfish then fluke, you can put together some decent catches if you target the area's bottom contours.

Better fluke fishing south of Double Creek can be had in the waters east of Gulf Point. Many anglers fish the shallows and the channel that winds east towards Long Beach Island. Though blue claw crabs can be quite thick off Gulf Point during the summer, patient anglers sticking to live killies and uneven bottoms generally do well. When fishing these and the previously discussed areas, think to yourself: "Where would I lie to ambush prey if I were a fluke?" and you'll usually locate fluke.

Chapter 3
Striped Bass

No fish that prowls New Jersey's coastline is as highly regarded as the striped bass. What makes the striped bass special to many anglers is it's a rare combination of size, strength, beauty and speed. Combine all these qualities with fine tasting flesh, and one can begin to understand why so many anglers pay homage to the "king" with late nights and early mornings.

Known as "striper," "rockfish," "squid hound," "linesider," "white bass," or "greenhead," the striped bass (Morone saxatilis) is a rather bulky fish that resembles a freshwater largemouth bass. A striper's upper back and relatively large head can range in color from a vibrant olive-green (hence the name "greenhead") to an inky black, while its flanks are a silvery-white emblazoned with seven to eight longitudinal dark stripes than run from the bass's gill plate to its tail. Occasionally, a very, very light crimson will run along the bass's sides. Stiff, thick spines comprise the bass's dorsal fin, which can easily inflict deep cuts. A striped bass's large, hard mouth can open as wide as the fish and contains fine, sandpaper-like teeth.

Within Barnegat Bay, striped bass range in size from 5-pound "shorts" or "schoolies" to 50-pound "cows." Save for an odd-ball catch like a drum, sting ray or shark, the largest fish an angler is likely to catch in Barnegat Bay is a striped bass. With striped bass believed to attain 150-pounds and New Jersey currently boasting the all-tackle world record at 78-pounds, 3-ounces, it's no wonder why so many anglers invest their time and money seeking the "king."

A Diverse Diet

A wide variety of baits appeal to striped bass — to say the least. Common striped bass baits include: bunker, crab, clams, squid, herring, mackerel, eels, mullet, minnows, sand bugs, snappers, sand eels, and sea worms. First-hand observation and experimentation should guide in selecting baits to target striped bass. When fishing for stripers with bait, gear your bait to match whatever organism or organisms the stripers are feeding on in the area you plan to fish, and you should have one piece of the puzzle

Seven to eight telltale stripes mark the striped bass, which is arguably the most highly regarded fish that swims New Jersey waters (opposite).

solved. Many anglers, for example, fish whole surf clams along the Island Beach and Long Beach Island surf after a strong storm because the turbulent waters will stir up clams and crush them in the surf. Stripers patrolling the surf, meanwhile, will feast on the clams, and the anglers using clams for bait will usually catch fish while everyone else gets skunked.

Bloodworms and sandworms are the dynamic duo of bass baits. Whether anglers drift them whole via a fish-finder rig or jig them with a leadhead jig, these two sea worms make excellent bass baits. Fat, lively worms that are about 10-inches long are the better worms to employ for bass. Anglers commonly use bloodworms or sandworms during the spring when the first bass appear in the bay. See page 59 for information regarding how to hook worms.

Just as fluke gravitate towards killies, bass have a few baitfish they prefer to feed on — bunker, herring and mullet. Once bunker and herring arrive during the spring, anglers begin either livelining them or cutting them up for bait. Mullet are particularly effective during the fall when these fish begin their migration and the bass feed on them.

Barnegat Bay's bunker population is usually in the 3 to 7-inch range, with the small "peanut" bunker being more predominant. Besides livelining, anglers fish fresh bunker steaks or chum with bunker and fish bunker chunks bluefish-style.

The bay's herring are typically in the 4 to 8-inch size range. Anglers usually stick to livelining these extremely soft-fleshed fish rather than cutting them up for bait.

The mullet which grace the bay range from small 4-inch "finger' mullet to foot long specimens one can catch on light tackle. Anglers frequently liveline mullet but also drift strips fashioned from the large specimens.

A fun and cheap way to acquire these three baitfish, is to catch them yourself. Tackle stores will sometimes carry these baitfish when the bass fishing is in full swing but for a consistent supply, the burden of acquiring theses baitfish falls on the angler. Bunker, herring and mullet can be found in many parts of the bay, particularly at Barnegat Inlet, Island Beach, Oyster Creek, Forked River, and the numerous small lagoons carved into the bay's western shores. Within these areas, one can catch these baitfish via a net, bunker snagger or with light tackle.

For netting, anglers employ seine nets when the baitfish are schooled close to shore. Small specimens occur near "seinable" areas while their larger brethren, which usually make better livelinees, tend to stay in deeper water or away from shore. To net the larger fish, and even small ones, quick-handed anglers use cast nets when the baitfish are near the surface. Daybreak and dusk seem to be the better times to find and net these baitfish at Island Beach and in the rivers.

When large baitfish are tightly schooled near the surface, many anglers snag a few via a weighted treble hook, commonly sold as a "bunker snagger." To use the snagger, simply cast the snagger into the baitfish school and quickly retrieve it through the school by giving the snagger quick, sharp jerks so the snagger's sharp treble hook snags a baitfish. Once snagged, many anglers let the baitfish swim around injured to catch a bass's eye. Others catch a few baitfish and either rig them for

Eels are perhaps the most popularly used bait among striped bass anglers. Though they may cause gray hairs with their slime and proficiency at

livelining or cut them up for bait.

A really fun way to catch herring and especially mullet is with light tackle. Bunker do not seem too interested in the doughballs herring and mullet will hit. One can catch herring and mullet via a handline baited with small doughballs molded to the hook's point. Chum with white bread and once you locate a school, bait an Eagle Claw salmon egg hook snelled to about 10-feet of 4-pound test with a doughball. Many times you will be unable to see these fish, because of their dark backs, but the doughball's movement when the fish hit it will give away their presence. You can generally catch herring throughout the summer in the bay's lagoons with doughballs while mullet seem to favor the late summer.

entangling themselves in nets and lines, anyone who has ever correctly fished live eels for bass will agree that eels are the perhaps the best bass bait, especially at night. Though anglers use eels from May to December, many experienced anglers tend to favor eels for bait during the fall.

Eels are easy to acquire through tackle stores, though they can be pricy at $4 a shot. Eels are also readily available to anglers who want to catch them with an eel trap or with light tackle. See Chapter 7 for information about locating and catching eels within Barnegat Bay.

The best eels for livelining in the bay are ones between 8 and 14-inches long. The small, 8-inch "shoelace" eels tend to get harassed

Striped bass respond to a variety of forage. Astute attention to bait and lures that match what the bass are feeding on will result in consistent catches of larger size bass like the one the author's brother took during the late spring (center).

more by snappers and large sea robins than bass. Some anglers really go ape and liveline 16-inch eels when targeting large bass during the fall.

Unlike many baits discussed thus far, eels are a very hardy species that do not require the delicate attention some live bait, such as herring, needs. Many anglers simply keep eels on ice with no water. Not only will the eels survive for quite awhile, but the cold will make them lethargic so the angler

squid, that are fished in strip form, use fresh bait which possesses a firm texture and fresh scent rather than some freezer-burned specimen turned yellow with age. Cutting a nice 10-inch long, $\frac{1}{2}$-inch wide strip from a fresh Atlantic squid is dynamite for stripers whether you fish it alone or use it with a bucktail. These strips are usually thicker and more durable on average than strips cut from smaller, foreign squid and if cut correctly, will flutter similar to a strip

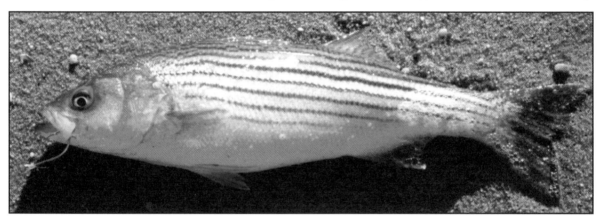

"Shorts" are fun to catch via light tackle but be sure to gently release them for the future. A 50-pound bass you may catch tomorrow might have been a small rascal that an angler in the distant past made an effort to release safely (center).

can easily handle them and prevent any slimy mayhem from erupting. Other anglers forgo the ice and place their eels on moist burlap, which substitutes as a grab rag to grasp an eel, or seaweed so the eels stay moist.

Though squid is the bay's universal bait, it's often overlooked by "serious" bass anglers. Considering that striped bass did not receive the name "squid hound" for nothing, squid is a viable option when fished in strips. The best squid to use for stripers is the native Atlantic variety that can be purchased at some tackle stores or jigged in New Jersey's inshore waters. This squid is not the 7-inch specimens fluke or weakfish anglers fashion strips from, but rather the 10 to 12-inch variety. With baits like

cut from a small, thinner fleshed squid.

Other popular baits for bass that find limited duty within the bay are: sand fleas, calico crabs and surf clams. Anglers who fish these "surf-oriented" baits typically concentrate their efforts in the inlet area or along the surf of Island Beach and Long Beach Island. Anglers typically fish these baits with a fish-finder rig.

Anglers fish sand fleas and calico crabs basically the same way tog anglers do. For sand fleas, simply use the whole flea. See page 98 for a detailed account on how to hook these rascals. For calico crabs, use whole small ones or half of a large crab. Hook the crab tog fishing-style (see page 98) but be sure to leave the crab's top shell intact rather than removing it.

When using surf clams, be sure to employ live ones, that are actually pulsating when you hook them, rather than the mushy frozen clams. When crabs are on the rampage, live surf clams really show their worth because the toughness of their flesh makes it hard for crabs to fleece the hook undetected. Frozen clams, on the other hand, are usually taken so stealthily that anglers are oblivious to the fact that they don't have any bait until they reel up. Also, live surf clams have a much stronger scent than the frozen variety.

Tackle and Techniques

Livelining baitfish is a potent way to consistently catch large striped bass. Rather than create an illusion with lures, liveliners forgo the smoke and mirrors and give bass the real McCoy. Bunker, herring and mullet are the chief baitfish anglers liveline for stripers. Snapper, spot and mackerel (both Atlantic and tinker) occasionally adorn a hook. Eels, which I did not forget, are perhaps the most popular livelinee and will be discussed shortly.

A good rod outfit to liveline baitfish starts with a stout, 6 to 7-foot stiff-tipped rod rated for 30 to 80-pound test. A conventional reel loaded with 20 to 30-pound test completes the outfit. Some anglers go as light as 15-pound test because the trick to livelining is to rig the baitfish so it swims naturally. Many anglers employ braided lines when livelining because they, unlike monofilament, don't stretch thus offering better sensitivity. Heavy line, such as 50-pound test, will weigh small

baitfish down and poor hook placement will mortally wound it.

Assuming you have a baitfish properly rigged and are fishing at a bass holding location, such as the inlet's north jetty, lightly toss the baitfish from your position, and let it swim near the rocks, ledge or other structure where a bass would likely hang to ambush its prey. Ladle out line to the baitfish, so it can swim about unhampered. Leave your reel in free spool and keep your line somewhat loose so a bass does not feel you when it grabs the baitfish.

Once a bass grabs your baitfish, the interesting part of the game begins: setting the hook. What makes setting the hook tricky is timing. Though most bass will grab your baitfish, run with it and then stop and eat it; others will grab it and drop it a few times before eventually consuming it. When you first feel a fish grab your baitfish, give the fish line. If you have been keeping your line loose, the fish will not detect you. Generally, a bass will take line after it grabs a baitfish because it's slinking off to eat it. After a few seconds, the line will usually stop, which is the bass turning the baitfish around to eat it. Now is when you engage your reel and get ready to strike. When your line begins to move once again and becomes tight, aim your rod tip downward at the fish and set the hook. Three or four strong strikes will firmly set the hook assuming your hook is sharp. Some anglers will purposely strike early, before the bass stops to swallow the baitfish, to avoid "gut hooking" the bass which saves the fish from significant injury.

Livelining may be tricky to master, but it is a very consistent way to catch bass. When you eventually catch a bass and wish to release it, revive it before release. The old heave-ho over the side will not usually cut it when trying to release a tired fish. Quickly remove the hook, trying not to handle the fish too much so you do not remove its natural slime coating. If gut hooked, cut your line and the hook will likely rust out. Gently place the bass in the water, and gently maneuver the bass through the water to stimulate water through its gills. After a few minutes, most bass will swim away.

The Eel Deal

If you want to significantly increase your odds of catching a bragging size bass, you should strongly consider investing in eels. Eels are to striped bass what hermit crabs are to tautog or shedder crabs are to weakfish.

Bassoholics typically liveline eels, similar to the way anglers liveline

Live Baitfish Rig and Hooking Techniques

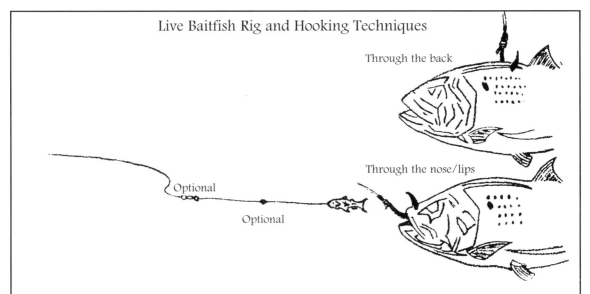

Through the back

Through the nose/lips

Optional

Optional

(Left) To liveline a bunker, herring or mullet, take a Mustad short shank bronze tuna hook and snell it directly to your line. Hook sizes range from 4/0, used for large bunker, to 6/0 for small herring. Make sure your hook is sharp because if you are not fishing with sharp hooks, you are missing the point in more ways than one! A bass's mouth is rather hard, not soft and supple as it may appear. As bass become larger, their mouths become bone-like which requires you to fish with very sharp hooks if you want to firmly anchor the hook in the bass's mouth.

(Optional) A few anglers use a swivel when livelining to prevent line twists. They will generally snell their hook to a 4 to 6-foot piece of 20 to 30-pound leader. They then connect the leader's other end to the swivel. When employing leader, remember the shorter the leader, the better control you will have over the baitfish, but a predator will be more likely to detect you. The longer the leader, the harder it is to control the baitfish, but a predator cannot detect you as easily.

(Optional) Some anglers like to employ weight to either bring the baitfish down to where the bass may be holding or to slow a baitfish down in a fast, turbulent current. Many livelinees, particularly eels, naturally seek the bottom so one does not have to usually use weight to get the baitfish down deeper. If you do want to get the baitfish deeper or slow it down, do not overweight the baitfish so it sinks straight to the bottom. Keep the weight light with fly fishing strip lead or a few splitshots about 10 to 15-inches above the baitfish.

(Right) You can hook a baitfish either forward of its dorsal fin or through its nose killie-style. The back hooking method is very popular. Slightly before the baitfish's dorsal fin, insert the hook no deeper than $1/4$ -inch into the baitfish's back. You want to insert the hook deep enough so the baitfish will not fly off the hook with a slight toss but is not mortally wounded where it does the dead man's float. Be especially careful when hooking herring because they are very delicate.

baitfish. Dead eels are usually either rigged on a tin squid and retrieved like a lure or are skinned and the skin is placed on a plug.

An excellent rod and reel combination for livelining eels is similar to the one employed for livelining baitfish: a stout, 6 to 7-foot stiff-tipped rod equipped with a conventional style reel spooled with 15 to 30-pound test. Many anglers favor dacron over monofilament for its sensitivity and superior hook setting properties because it, unlike monofilament, does not stretch. Whatever type line you choose to employ, use at least 20-pound test if you are fishing around rocks, pilings or any other abrasive structure that can chafe your line. Considering a large bass can maybe break your line outright, a small nick will only make a bass's escape much easier.

When livelining an eel at a potential hot spot, such as the inlet's north jetty, encourage the eel to swim where a bass would likely be lurking to ambush prey. This means that if you are livelining at a jetty, rather than heaving the eel 50-yards from the jetty, let the eel swim amongst the rocks. Old salts don't call striped bass "rockfish" for nothing. If drifting eels with a sinker, keep the sinker and eel close to but not dragging along the bottom. Let the rig's sinker hit bottom, and then reel up a turn or two. Periodically re-establish contact with the bottom.

Assuming you are livelining in a bass holding area and put in your time, a bass will sooner or later grab your eel. Bass will grab an eel almost exactly the way they grab a bunker, and

Live Eel Rig and Hooking Techniques

Through both lips

Optional

4'

Through the mouth and out the eye

(Left) For an eel livelining rig, snell a Mustad #9174 hook to a 4 to 5-foot piece of 30 to 50-pound leader material. Hooks range in size from a 3/0, for large eels and large bass, to 6/0 for small eels and small bass. Tie the leader's other end to the swivel, which prevents your line from twisting, with a clinch knot. With leader lengths, remember the shorter the leader is the easier it will be to control the eel. With longer leader, you lose some control but the presentation appears more natural because the bait can maneuver more freely.

(Optional) Some anglers when faced with a deep location or a fast current, will attach a 1 to 4-ounce egg or rubber core sinker just above the swivel to keep the eel close to, but not dragging along, the bottom.

(Right) When livelining, anglers typically hook an eel either through the lips killie-style or through the bottom jaw and out the eye socket. Others will simply hook the eel through both eyes, Whichever way you choose, be sure the eel is firmly attached to the hook. by making sure the hook's barb is clearly visible. If not visible, push the hook through because eels have a habit of wriggling off hooks.

When attempting to hook eels so they stay on the hook without creating a big mess, use either a scouring pad, fish-gripper glove or burlap grab rag to firmly grasp the eel and hook it. Bare-handed attempts are usually frustratingly futile. Keeping the eels on ice or in ice water to slow the eel's metabolism down enough so you can easily grab them with a rag and hook them without getting your line, tackle and self slimed.

anglers should give line and set the hook the same way they would if fishing for bass with a bunker.

After you have landed or missed a bass, do not be so quick to toss your mauled eel aside. Whether it's because a spat-out eel appears as an easy kill or emits a strong scent in the water, stripers often show a preference for previously used eels that still show signs of life. Using mauled eels also saves you bait!

For eels that gave up the ghost, you can rig them with a tin squid and work them like a lure. Commonly known as "rigged eels," an eel rigged to a tin squid swims in a sweeping "S" pattern that gives anglers the flexibility of a lure wedded to an eel's look and scent.

To rig a dead eel to a tin squid, simply snell a 5/0 short shank tuna hook to a 12-inch piece of 50-pound dacron line. Attach the line's bare end to a rigging needle, slide the rigging needle through the eel's anus and out its mouth. Pull the line so the snelled hook goes into the eel's anus so only the hook's bend and point are visible.

The tin squid's hook is then inserted through the center of the eel's head. The line, which should be jutting through the eel's mouth is tied to the tin squid's eye. Your main line is then tied to the hole on the tin squid's lip.

If properly centered on the tin squid, the eel should "swim" in a sweeping "S" pattern when retrieved. It is smart to test your eel in calm water, preferably in a lagoon from a dock or pier so you can see if your eel will swim straight in the sweeping "S" pattern before you try casting it off a jetty in darkness. When fishing, retrieve the eel slowly, so it lazily "swims" through the water.

Other Rigs

The basic fish-finder rig is one of the more popularly used rigs for striped bass, especially when using cut or strip baits. Anglers typically fish squid strips, sea worms, clams, calico crabs, cut bunker, and sand bugs via this rig, which boasts no hardware separating the angler from whatever grabs the bait.

Fashion the fish-finder rig the same way you would for fluke but employ a size 4/0 or 5/0 #92642 Mustad sliced shank hook or 1/0 to 3/0 beak style hook instead of an English wide gap hook and 40-pound leader material instead of 15-pound leader when fishing most baits.

Many anglers use a fish-finder rig when fishing clams because this bait often attracts less desirable species, such as sea robins and crabs. The fish-finder rig lets the angler easily detect a bait stealer while other type rigs, such as the 3-way swivel rig, give bait stealers more latitude.

The fish-finder rig also works well when drifting cut bunker, a striper favorite. Anglers typically cut bunker into 2 to 4-inch wide steaks with the bunker's head being widely considered the best piece one can fish. Fish a single steak and hook it once through the "meaty" section so it is firmly anchored to the hook. Hook the head by inserting the hook behind the gills and through the meat.

Recently caught bunker, kept fresh on ice, is the better to employ than frozen or rancid bunker. For one, fresh bunker is much firmer which causes it to stay on the hook better. Secondly, fresh bunker seems to emit a fresher, oilier scent through the water that attracts bass. If you obtain bunker through your own skill with a bunker snagger, try drifting a steak cut from a snagged bunker in the immediate area. Chances are good that there are bass feeding on the same bunker school you are picking at with your snagger.

A Lure Thing

Many anglers forgo the hassle bait presents and exercise their arms with artificial lures. Before bait anglers scoff at a lure caster's chances at catching a bragging size bass, they should remember it was a Rebel plug that Al McReynolds used to catch the world record striped bass.

Striped bass respond to a wide variety of lures, ranging from plugs to poppers. Proper lure selection involves matching the lure to the baitfish the bass are feeding on and manipulating the lure to mimic these baitfish. Very rarely will striped bass strike any lure that possesses some flash and movement like bluefish will do at times. Striped bass will aggressively hit a lure, but the lure usually closely resembles the area's predominant baitfish in appearance and movement.

Though each type lure requires its own unique presentation for an angler to properly mimic a baitfish, anglers generally work lures for bass slowly. Striped bass tend to respond to lures when worked slow, weakfish-style rather than at mach 1 bluefish or Spanish mackerel style.

Plugs are perhaps the most popular lure group retrieved for bass. Plugs come in an array of colors and sizes. On the bay scene, the better known plugs are: Bombers, MirrOlures, Rebels, Cordell Redfins, Creek Chub

Sea Worm Rig and Hooking Technique

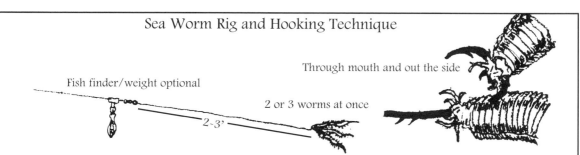

Through mouth and out the side

Fish finder/weight optional

2 or 3 worms at once

2-3'

(Left) To fish sea worms with a fish-finder rig, the better hooks to employ are: 3/0 to 7/0 short shank tuna hooks, 1/0 to 3/0 beak style hooks or a #92642 Mustad sliced shank hook. The better worms to employ are lively, long ones about 8 to 10-inches long. Hook a worm by inserting the hook into the worm's mouth and bringing the hook back through the worm's head. The distance between the worm's mouth and where the hook juts from the worm should be about 3/8 of an inch.

(Optional) When fishing worms with the fish-finder rig, bounce the sinker along the bottom as you would for fluke. The sinker bouncing will make the worms appear as live worms swimming through the water (left). You can also employ no weight and drift the worms with the current through a potential bass-holding area, similar to the way a trout angler would drift a mealy worm in a stream for a brook trout.

(Right) Make sure you center the worm on the hook so it will flow smoothly in the water. If you choose to go the multiple worm route, with many anglers fishing three or four worms at once, make sure the worms are hooked so they flow naturally like hair rather than a glob of putty. Substituting a Berkley Power Sandworm for a real worm will often add length and stability to a pair or trio of fragile sea worms.

pikies, Rapalas, Stan Gibbs swimmers, and Zara Spooks.

With colors, anglers will find silver-sided plugs with a black, blue or green back work best in the bay when mullet, herring, bunker, or snappers are present and bass are feeding on them. The all black Bomber plug is a very effective plug to use at night.

Five-inch Cordell Redfins, 5½-inch Rebels, 15A and 16A Bombers are a few particularly good plugs to work when probing the bay for legal-sized bass. For schoolie bass, try using a F-10 or F-20 Rebel, or a 14A Bomber. Try casting these size plugs at locations where shallow water teeming with baitfish drops sharply to deep or swiftly moving water, around rock piles or near deep holes. The flats behind Island Beach, the inlet's western mouth, the inlet's two jetties, Oyster Creek's eastern mouth, and the Mud Channel are a few places in the bay where plugs account for bass.

Savvy anglers match their plug to the area's prevalent baitfish and retrieve the lure slowly and let the plug's built-in action come to the forefront.

Teasers come into play when small baitfish prevail in an area, and anglers attempt to mimic the small baitfish. Teasers are small lures tied before a main lure and usually take the form of unweighted bucktails, saltwater flies, twister tail grubs, plastic worms, pork rind strips, shad darts or Fin-S-Fish.

Anglers typically tie a teaser to a 5 to 8-inch piece of 30-pound stiff monofilament leader and then tie the leader about 2-feet before the lure, which is usually a plug. When the angler retrieves the plug and teaser combination, the plug appears as a large baitfish pursuing the teaser. This sight often causes a bass to strike at either of the two. Many times you will find the larger bass taken on the smaller teaser rather than on the larger plug.

Besides plugs, an old stand-by for lure casters who ply the brine off the northeast is the bucktail. Bucktails are particularly good to use when targeting the numerous sub-legal shorts that populate the bay. Ideal sizes for bucktails range from 1/8-ounce to 1¾-ounces with pure white being a popular color.

Good bucktail sweeteners for striped bass include: Mann jelly worms, Uncle Josh bass strip, thin bunker fillets, mullet fillets, mackerel strips, or long, thin squid strips.

When fishing for striped bass with a bucktail, retrieve the bucktail basically the same way you would for fluke — slow hops along the bottom. Keep your retrieve slow and mix in short bounces, that keep the bucktail close to the bottom, with quick lifts, that cause the bucktail to dart like a wary baitfish.

Bucktails are particularly effective when bounced at the bay's calmer backwater areas, such as the flats behind Island Beach, or tight to structure, such as the Mantoloking Bridge or the inlet's jetties rock piles.

Within recent years, the Fin-S-Fish has gained popularity with bucktail bouncers. Anglers work this soft plastic fish with a jig head exactly as they would a bucktail. When targeting bass, savvy anglers dip their Fin-S-Fish in bunker oil to mask the lure's plastic aroma with the bunker scent bass love.

Striped bass, like bluefish, typically respond to metals because these silvery, slender lures closely resemble the sand eels, spearing, peanut bunker, and finger mullet bass commonly devour. Metals are especially good to fish whenever you happen upon a school of tailor blues shredding a hapless baitfish school. A metal, such as a Need-L-Eel, allows you to drop below the bluefish action, which is usually

do a good job imitating sand eels. Use a 007 size when small sand eels abound while resorting to a A27 when larger specimens are prevalent. In addition, 1/8 to ³/₄-ounce Bridgeport jigs and similar sized Need-L-Eels imitate small sand eels, spearing and rainfish. Crippled Herring lures, in the ³/₄-ounce to 1¹/₂-ounce size are a good match for small herring, peanut bunker, finger mullet, and small snappers.

confined to the surface, and work the bottom tier which is usually occupied by striped bass. On several occasions while spot casting for bluefish in the bay, I have witnessed a fishing companion bring a blue to the boat only to see a human-sized striped bass materialize from the dark water, see the boat and then quickly disappear with a large flash!

Savvy lure casters usually confine their metal use to match local baitfish or as a no-frills method to casting against a howling wind. Ava jigs

When using metals for bass, you generally want to try a slow retrieve along the bottom and let the metal's built-in action do the initial work. The trick with working metals properly for bass, is imitate a baitfish's movements to fool a bass. Thus, while you want to employ a metal's built-in action initially, you will want to experiment with mixing up your retrieve to see what movement pattern best matches the baitfish. When using small ¹/₄-ounce Bridgeport jigs, for example, try a slow retrieve that's generously altered with

In essence, striped bass are nocturnal creatures. Thus, try late night fishing trips if you want to consistently score with striped bass (center).

quick jigs causing the lure to dart just as a spearing would if it wanted to keep pace with its school.

In addition to metals, many anglers employ poppers for bass because they enjoy the excitement poppers usually offer — the sight of a bass erupting from the water's surface as it attempts to engulf the lure. Poppers are surface lures designed to imitate injured or fleeing baitfish and are generally employed in areas where bass are known to feed at the surface. Poppers are especially effective during the mid-September to mid-October time frame when mullet are usually tightly schooled and migrating.

Common poppers used around Barnegat Bay are the: Atom, Creek Chub, and Gibbs. The chrome, black-backed Striper Strike in the 5-inch size is the popper of choice when mullet are present while a 4 or 5-inch blue/white Atom works just as well for the bay's bass as it does for its bluefish.

When working poppers for bass, the same general "popper steps" used for bluefishing apply: jerk and stop the popper to mimic the irregular motion of a fleeing or injured baitfish. You do, however, want to work a popper s-l-o-w-e-r than you would for a bluefish. When a bass grabs your popper, strike when you feel the bass's weight rather than when you visually see the bass grab the popper. Otherwise, you will usually either strike at water or feebly hook the bass so only Neptune's good graces will allow you to get the thing in without it busting off.

Where and When

Striped bass occur from the Gulf of the St. Lawrence River in Canada to Florida but are most commonly found between Cape Cod, Massachusetts and Cape Hatteras, North Carolina. They also occur along the Pacific Coast of North America from British Columbia to Mexico. Though they can be found over any type of bottom, striped bass prefer areas where they can enjoy a steady food supply and set ambushes for their unsuspecting prey. Such areas include: rockpiles, sharp drop-offs, channel edges, bridge pilings, sedge banks, salt marsh islands, creek mouths, river mouths, and sand bars.

Technically, fishing for striped bass can be a year 'round endeavor because bass will maintain a resident population at certain areas of New Jersey during the winter. For the most part, productive bass fishing will generally start around early April along the surf or in the rivers. Fishing for bass usually becomes great during the late May to mid-June time frame when the water temperature hangs in the mid-50 degree range. Once the water heats up under the hot July sun, stripers are not as conspicuous as they were when the water was in the mid-50s. Stalwart anglers plying stagnant July and August nights will continue to enjoy catching bass. Once the water begins to cool around mid-October, bass fishing starts to pick up once again with the late October to late November time frame typically offering the year's best bass fishing opportunities.

To consistently catch striped bass, one has to consider the time of day, tide, moon phase, and wind because these four items play a hand in where bass will be located. Bass are essentially a nocturnal species. It is under the cloak of darkness that they invade shallow water locations to feast on baitfish. It is then, when you should be wetting your line.

Tides also figure into the puzzle. Bass will typically invade shallow water during an incoming tide that floods the shallow water location. During the night, bass will venture into relatively shallow water to hunt baitfish. Keep this in mind if you are near the flats behind Island Beach and the sun sets. During a receding tide, bass will typically hang just on the outskirts of shallow water locations, creek mouths or inlet mouths and wait for the tide to drag baitfish from the shallows to the deeper water.

The moon comes into play because it affects the tide. During a new or full moon, expect extremely low, low tides and extremely high, high tides. During these moons, shallows, that previously offered baitfish refuge, become much deeper at high tide so predators can raid the area while at low tide some baitfish refuges become terra firma and baitfish are forced to venture into a bass's backyard. Couple the strong tides that occur during a new moon with the dark, moonless night, and you have all the makings for a good bass night.

Upper Bay

When considering the upper bay for striped bass, the Point Pleasant Canal is a noteworthy location that attracts many anglers. From the bay's perspective, the canal's mouth is a good place to look for bass because it is here where the canal's strong currents wash baitfish into the bay. Anglers should concentrate their efforts at dusk and prior to daybreak when the bass are active and not scared by the sun or boat traffic. Bucktails, Fin-S-Fish and Redfins are good lures to try while finger mullet, sandworms and bunker steaks are good baits to drift at the mouth.

Southward, anglers catch bass along the channel that winds past Herring Island. Most anglers have better luck here at dusk, night or daybreak when there is minimal boat traffic. Herring Island's eastern shores, along the channel's ledge usually provide bass to lure casters and bait drifters. One particularly good combination that works well for bass here is a $^1/_2$-ounce plain leadhead with a long, fat bloodworm or sandworm threaded on it.

Anglers drifting worms, livelining baitfish and casting lures for weakfish at the Mantoloking Bridge will typically catch striped bass. Most find bass tight near the bridge's foundation and along the channel that winds beneath the bridge at night or daybreak. This is when baitfish are stirring and most boaters aren't.

Although anglers can occasionally find striped bass at the Sedge Islands west of Chadwick Beach, the Route 37 Bridge, Pelican Island and within the Toms River, the better bass fishing, perhaps in the bay, is to be had at the flats located a tad southeast from the Toms River. These flats, which extend from Seaside Park all the way south to Oyster Creek Channel provide acres of untrodden water that teem with bass forage. Because the flats are

Did You Know?

A 50-pound striped bass is approximately 18 years old while a 25-pound specimen is around 12 years old.

Reeves Cove *The Flats* *Tices Shoals*

The flats that lie west of Island Beach are a prime area to find striped bass. Reeves Cove and Tices Shoals are two popular areas to fish at the flats (center). Do not, however, overlook the edges of backwater salt marshes because these areas usually host bass (top left).

relatively close to the inlet, a prime entry and exit point for striped bass, anglers will find striped bass holding court at the flats throughout the summer. Though the bass you are likely to encounter at the flats are not gigantic monsters, you will tangle with legal sized bass and plenty of small bass if you explore the flats.

From a striped bass fishing perspective, the flats are bracketed by two fixtures: Reeves Cove to the north and Tices Shoals to the south. What makes Reeves Cove special is it possesses a mean depth of 6-feet while extending into the flat which is about 2-feet deep.

After dark, anglers can typically find stripers prowling the cove's edges and the flats itself seeking prey.

From Reeves Cove south to Tices Shoals, anglers can cast lures or work bait for bass along the flats' outer ledge. If you own a small boat, try venturing into the flats and look for the small holes and bottom contours where striped bass, bluefish and weakfish frequently dine. Poking along the flats is a good alternative to Reeves Cove or Tices Shoals when these popular spots are clogged with boats.

Widely regarded as the mecca for the bay's weakfish, Tices Shoals also

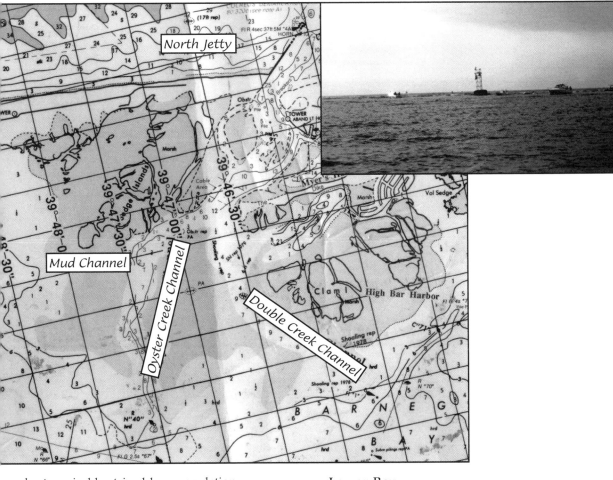

North Jetty

Mud Channel

Oyster Creek Channel

Double Creek Channel

hosts a sizable striped bass population. Similar to bluefish and weakfish, striped bass prowl the edges of the shoal when hunting baitfish, and many anglers catch bass here while chumming for bluefish or weakfish. Bomber or Rebel-style plugs are good to work at the Shoals when there are no weeds present in the water. When there are weeds, try a bucktail, sea worm and leadhead combination or a Fin-S-Fish. Try a trip to the shoals just before nightfall or at daybreak. If the water is quiet and you're patient, you will see Tices Shoals in a different light.

Lower Bay

When considering the lower bay for striped bass, the first area that warrants investigation is the Oyster Creek area which encompasses Oyster Creek and Oyster Creek Channel. With a nuclear power plant on its shores, Oyster Creek is one of the first locations within the bay to become productive for bass. Around mid to late April, anglers can often find bass in Oyster Creek feeding on the baitfish, attracted by the warm water discharged from the Oyster Creek Nuclear Power Plant. Though the bass one is likely to catch at Oyster Creek rarely reach, at least in my

Anglers can typically find good bass fishing within close proximity to Barnegat Inlet, at such locations as the Mud Channel, Oyster Creek Channel, Double Creek Channel, and at the inlet itself (center). The inlet's venerable north jetty, whose bell tower still defies the Atlantic is a particularly popular area for anglers looking for a trophy size bass (top right).

experience, colossal size, they are big enough to make fishing with light tackle interesting.

Oyster Creek Channel, which extends east from Oyster Creek to the inlet, is a comfortable 10 to 12-feet deep but immediately outside the channel the water rises to a 3-foot deep eel grass flat. Look for bass to hang near the ledge formed where the channel borders this flat. Bass love to patrol such ledges looking for baitfish. During the spring and early summer, one can often find striped bass lurking beneath the bluefish schools that commonly maraud the channel.

The action in Oyster Creek Channel tends to heat up as you near the inlet. A good place to investigate is where the channel runs adjacent to Island Beach. Here, a few creeks brimming with baitfish activity empty into the channel, and striped bass merely hang downtide from where the happy 2-foot deep creek drops into the swirling, chaotic 15-foot plus deep channel. This is a popular location to liveline eels and other baitfish. During an incoming tide, if the water depth permits and it's dark, bass will venture very close to these creeks.

A few years ago my brother, myself and our friend Chris were casting plugs around this Oyster Creek/Barnegat Inlet area at dusk before the onset of a storm. Though the incoming tide was raging, we sought to get closer to Island Beach to work our lures. Shortly thereafter, our boat's engine failed, and it became very apparent that the tide was going to sweep us over the, by now submerged, salt marsh into a large tidewater pond that was filling with the incoming water

about 100 yards inland. As my brother worked to start the engine, my friend Chris and I jumped onto the once dry marsh bank, that was by now under 3-feet of water, to prevent the boat from being pushed back to this pond. As Chris and I were flexing our muscles, I noticed a school of five large striped bass swim with the raging tide back to this pond. One point I gathered from this experience is that bass will venture into seemingly inaccessible locations to hunt baitfish. Keep this in mind when you see birds hitting a baitfish school on the flats, and your initial thought is snappers rather than 40-inch bass are the perpetrators.

North of Oyster Creek Channel lies the fabled Mud Channel which is a popular location for striped bass. The Mud Channel is a good place to work plugs or bucktails at daybreak or liveline eels at night. This channel is usually a productive area during an incoming tide when the incoming water from the nearby inlet stirs up the bass.

Not to far from the Mud Channel lies Barnegat Inlet, which is one location where anglers have a good shot at landing a cow bass throughout the year. From a bass fishing perspective, one location in the inlet that sticks out is the large sand bar the inlet's main channel snakes around en route to the ocean. Anglers typically work the edges of this channel with plugs, bucktails or rigged eels.

A stone's throw south from here, is a potent bass location: the sod shores of Long Beach Island. Here, the water is very deep a scant few feet from the banks. Anglers typically liveline baitfish towards the bank or work lures along the bank's ledge.

If we continue to follow the inlet's main channel towards the ocean, anglers typically fish bass at the small channel that branches off from the main channel and runs behind the large sand bar. Though this channel tends to attract a lot of shorts, you never know.

East from the sand bar are two prominent fixtures for bass anglers: the north and south jetty. It is from these two jetties that anglers have a good chance at catching a large bass, especially if they concentrate their efforts at night or daybreak.

The inlet's south jetty, which was completed several years ago, possesses many good pockets along its stretch for anglers to work for bass. Anglers fishing from the bulkheads at Andy's and in front of the lighthouse do well with bass at night. Similar to fishing the north jetty, try livelining baitfish or casting lures close to the jetty's rocks where the bass are likely to be prowling. Because the waters near the south jetty are shallower than the north jetty's waters, anglers typically fish the jetty for bass at night when the bass are most likely to enter the shallow water. One good point about the south jetty is anglers can walk south from the jetty and fish Long Beach Island's surf. The old south jetty and "the mast," are a few landmarks that offer great bass opportunities from a surf fishing perspective.

Across from the south jetty lies the north jetty which is usually flocked with anglers casting from its rocks and campers parked near its base. When fishing the jetty's bay side from a boat, you will usually have to worry about boat wake washing you into the rocks. When there is excessive boat traffic, fish either the jetty's tip or its ocean side where the water is calm. Many anglers anchor near the jetty's ocean side and cast, liveline or chum the jetty without worrying about the jetty's treacherous rocks. A good aspect about fishing both jetties, is if bass give you the cold shoulder, you can salvage the day by fishing for tautog.

For anglers sticking to the bay, Double Creek Channel is a popular bass location for liveliners. Boasting a close proximity to Barnegat Inlet, deep holes and relatively light boat traffic, anglers fishing Double Creek often enjoy good bass fishing. Areas to investigate in the channel are the channel's eastern mouth, where it meets the inlet, and the edge of the large sand bar as you follow the channel west into the bay, and off Clam Island. These are a few locations that offer some hot bass fishing, especially if you prefer to liveline eels and do your livelining during a moving tide.

With some investigatory work, anglers fishing Barnegat Bay will literally find striped bass lurking right beneath their nose.

Chapter 4
Weakfish

Possessing a chrome-like glow accentuated with purple speckles and splashes of yellow, the weakfish is one of the more attractive if not tastiest fish that roams Barnegat Bay. Every summer and fall, anglers try their hand at weakfishing and the number of boats that grace the bay's inlet, channels and flats testify to their sincerity.

Also known as "sea trout," "summer trout," or by its Native American moniker "squeteague," the weakfish (Cynoscion regalis), derives its name from its soft mouth and weak jaw that anglers often tear when hooking and fighting it. Weakfish are easily distinguished from other bay species by their purplish upper body that quickly fades into numerous purple speckles that stay above the fish's pectoral fin. A weakfish's head and body is a brilliant chrome-like silver or silvery-white with its pelvic, anal and pectoral fins a brilliant yellow.

Weakfish can grow as large as 25-pounds. Within Barnegat Bay, weakfish average in the 2 to 5-pound range. Within recent years, small, sub-legal sized weakfish, called "spikes," around a foot long have been especially prevalent within the bay. Eight pound and larger weakfish, called "tiderunners," also reside within the bay and are not uncommon, though not frequent catches. Seemingly every summer, some lucky angler manages to bag a 10-pound plus tiderunner while fishing Barnegat Bay.

Squeteague Sustenance

Weakfish are often very finicky when it comes to what and when they prefer to eat. Sometimes, weakfish only respond to fresh shedder crab baits while on other occasions, they will attack anything including a freezer burned squid head jigged via a balding bucktail. Weakfish will often make one go and scratch their head.

All things being equal, the baits chiefly favored by weakfish are: shedder crab, shrimp, spot (lafayettes), sea worms, and squid. If you are really bent on catching weakfish, especially ones of tiderunner proportions, you should consider investing in live bait. There are many areas within Barnegat Bay that host weakfish prey making it relatively

True to their name "sea trout," weakfish are intricately colored fish that resemble freshwater trout. They respond well to both bait and lures and are one of the bay's more popularly sought species (opposite).

easy and enjoyable to collect fresh bait. A good point about collecting your own bait, besides the freshness factor, is weakfish will often feed exclusively upon local forage which often means the weakfish prowling the general "bait collecting" area are feeding upon the organisms that you are collecting for bait. "Matching the hatch" becomes a no-brainer.

Shedder crab is considered by many weakfish anglers to be the best bait one can use for weaks. Also known as "peelers," or "shedders," shedder crabs are in actuality blue claws that are on the verge of discarding their hard shell in favor of a fully formed soft one lying beneath it. Anglers typically fish shedder crabs in small chunks within a shrimp chum slick or, more commonly, jigged with bucktails or plain leadheads.

One way to distinguish a shedder from a regular hardshell crab is to gently push on the crab's large, flat back flippers. If the crab is a shedder, its flippers will not be flat and thin as normal, but will rather be thicker and give to a little pressure. You can also try pressing the white shell area underneath the crab's top shell. If you have a shedder, the shell will give a little or actually split with some pressure, exposing the newly formed shell underneath.

Most anglers prepare shedders for bait as if they were trying to peel the crab for food. Initially, they cut off the crab's claws, to avoid getting pinched, and then pop the crab's apron on its underbelly. They then carefully remove the crab's top shell just as if they were going to eat it and cut the crab similar to the way one would a green crab for tautog. They remove the legs and

flippers with a snip from a scissor. They then cut the shedder in half, and following the lines on the crab's underbelly, cut each half into five pieces. Some anglers crack the crab's claws and use the claw meat threaded on a jig head.

The two most efficient methods to catch shedders is to either seine them, around the bay's marshes and eel grass "flats," or scrape them off pilings and bulkheads. Many use the standard baitfish style seine net to catch baitfish and shrimp while trying for shedders. If shedders are the focus, and cleaning the net of baitfish, shrimp and other assorted creatures is undesirable, try seining with a net that possesses a wider mesh, such as 2-inches. Smaller organisms will pass through the net while crabs and larger characters will not. When seining for shedders, sharpies will add heavier weights to the bottom of their seine nets so the net drags along the bottom, where the crabs are, rather than floating an inch or two above it and passing over most crabs.

Within the bay, the better areas to seine are eel grass shallows, the flats west of Island Beach, Clam Island, and the many small beaches and seinable areas that dot the bay's western shores from the Barnegat Public Docks to the Beaver Dam Creek. Particularly good areas in this expanse include: the Berkley Island County Park, the Sedge Islands and Cattus Island County Park.

Scraping for shedders involves scouting around bulkheading, piers, bridges, and marsh banks during the early morning or at night and scraping them off a structure with a scoop net. Stick to bridges, bulkheadings or other scrapeable structures, especially if they are near a river or creek mouth where blue claws generally thrive.

Readers should note that fresh, live shedder crabs are preferred as bait; frozen shedders are often mushy and do not seem to pack the same punch as live shedder crabs.

In the absence of shedder crabs, anglers will often turn to softshell blue claws. Softshells are frequently used when drifting or chumming because their extremely soft flesh is hard to keep intact on a bucktail or

as prevalent as grass shrimp. Anglers chum with live or freshly dead shrimp because if weakfish happen to be finicky, which they invariable become, live shrimp will always get them interested whereas frozen shrimp do not pack the same punch. Ergo, procuring your own shrimp becomes important if you want to consistently catch a lot of weakfish. Catching both varieties of shrimp via a scoop net or seine net is fun and easy.

Scraping for shrimp with a

other "castable" hook. Anglers using softshells will usually give the softshell chunks a few drops of shedder crab oil to impart the weakfish attracting scent unique to shedder crabs onto the chunk.

Shrimp rival shedder crab as the top weakfish bait. Many anglers use the small crustaceans as both chum and bait because shrimp are a weakfish staple. Barnegat Bay is loaded with two types of shrimp: grass shrimp and sand shrimp. Anglers can usually find grass shrimp in abundance at the bay's eel grass flats, marshes and attached to any barnacle-encrusted or algae covered structure, particularly bulkheadings. Sand shrimp, as their name implies, reside at the bay's sand bars and other sandy bottomed areas and do not seem

scoop net involves acquiring a fine mesh net from a tackle store or catalog. Nets with a mesh about 1/8-inch are good for scraping shrimp because the net's small mesh will prevent any shrimp from passing through. You then want to locate a "scrapeable" area where shrimp thrive, such as pilings, bulkheads, salt marsh banks, brackish creeks, and lagoons. The Barnegat Township Public Docks and the local bulkheading are two good places to scrape for grass shrimp while the venerable bulkheading that lines Island Beach's southern extreme in the inlet, is a good place to explore for sand shrimp.

To scrape shrimp off a structure, extend your scoop net down near the structure's base and work the

The weakfish is among the bay's more brilliantly colored fish. Note the fish's relatively large, powerful tail which makes it a fast swimmer and tenacious opponent. Closely examining the weakfish's mouth shows the weak jaw which gives the fish its name (center).

net towards the surface at a moderate rate. Scrape fast enough so shrimp can't avoid your net with a quick squirt, but slow enough so you don't crush any against the structure or tear your net against barnacles. Depending on the length of the piling and the length of your arm, you may want to lengthen your net with a pole before attempting to scrape for shrimp.

Seining, similar to scraping, involves a net with a fine mesh of about 1/8-inch wide. You basically seine for shrimp the same way you would for spearing, the only difference is you want to keep the seine net hugging the bottom. If the net's bottom weights are riding off the bottom even slightly, you will pass over most shrimp. Avoid this by adding heavier weights along the net's entire bottom.

When considering areas to seine for shrimp, remember that weakfish favor certain areas because these areas play host to organisms that wet a weakfish's palate. Ergo, areas that are good for weakfishing are typically loaded with grass and sand shrimp. Conklin Island, Clam Island, Tices Shoals, the flats behind Island Beach, and the shallows along Long Beach Island's western side are a few good areas to seine for grass shrimp. Tices Shoals, the sand flat the lies to the north of Double Creek Channel, and the large sand bar in the inlet are additional areas to try seining for sand shrimp.

Once caught, shrimp will keep in a 5-gallon bucket filled with bay water. To keep the shrimp alive, keep them shaded and occasionally freshen the bucket's saltwater. Others rather avoid the hassle and place their shrimp in a small cooler atop a damp rag or eel grass mat that lies above crushed ice.

If seining or scraping doesn't tickle your fancy, you can occasionally purchase fresh shrimp at some area tackle stores or through local bait netters, if you can locate them. Note that the majority of shrimp found at tackle stores is the frozen variety.

After a successful trip with many fond memories, by no means give any left-over shrimp the heave-ho because you can still use them to add bulk to a chum slick or scent to lures. Freezing the remainder in a plastic sandwich bag will keep leftover shrimp in good condition if the bag is airtight. Shrimp have a bad habit of getting freezer burn if not stored airtight or kept longer than about a month. While they will not be as potent as live grass shrimp, frozen ones can add bulk to a chum slick. Alternatively, you can fill an olive jar halfway with leftover shrimp, mash them up and reserve the juice as an attractant for lures. Shrimp juice will keep well for a few weeks in a small air-tight jar for application to bucktails or Fin-S-Fish.

Though shedder crab and shrimp are the more popularly used baits for weakfish of all sizes, live spot (lafayette) is a primary bait aimed at catching tiderunner weakfish. Spots resemble croakers in general appearance. They possess a rounded head and thick body which ranges in color from a light gray to a steel blue. A prominent black spot, located just behind the lafayette's pectoral fin, gives the fish its nickname "spot." Within the bay, spots range from about 3-inches to 8-inches.

Though one can fashion strip baits from a spot, and either drift them with a bottom oriented rig or bounce them on a bucktail, spots are most potent when livelined. Anglers typically liveline spot with a simple livelining rig, that is very similar to the one striped bass anglers liveline herring or an eel with. They primarily target locations that can potentially hold tiderunner weakfish, such as the Mantoloking Bridge, Tices Shoals, Oyster Creek Channel, Double Creek Channel, Meyers Hole and the inlet's two jetties.

Though area tackle stores rarely carry live spots, catching them with light tackle is fun and not too time consuming. Employing a simple fish-finder or hi-lo rig, tied with #6 or #8 bait holder hooks and baited with either small 1-inch bloodworm pieces or similar sized squid strips or hardshell clam bits is a good way to catch spot. Rocks, piers, channel ledges, lagoons, river mouths, and creek mouths are the primary areas where one can usually find spots. The Beaver Dam Creek, the Metedeconk River, the Toms River, Cedar Creek, Stouts Creek, and the numerous residential lagoons that lie south of the Toms River usually all hold spot some time during the summer.

Though spots are more durable than spearing, they are not as hardy as a killie. Thus, some attention on your part is required if you're entertaining the idea of fishing a somewhat lively bait. Keep spots in a live well, killie car, Flo-troll, or anywhere where they can enjoy a constant supply of freshly oxygenated water.

Sea worms, both bloodworms and sandworms, make excellent weakfish bait with anglers typically drifting them along the bottom with a bottom rig or leadhead jig. Whole sandworms are the preferred bait and are usually drifted with a fish-finder rig. Weakfish anglers typically hook a worm through the head the same way a bass angler hooks sea worms, but weakfish anglers usually fish a single worm rather than a duo or trio.

Besides drifting, jigging a worm with a leadhead jig is another popular way to fish sea worms for weakfish. Most anglers cut a worm into 5-inch pieces and then thread a piece on a bucktail or plain leadhead jig. Though placing a whole 10-inch worm on a bucktail usually results in a spike snipping the worm in half, savvy anglers will fish a whole worm on a jig to grab a large weakfish's eye because larger baits sometimes make a difference. When it's evident, by the worm's severed body, that spikes abound, anglers will shorten the worm.

One dilemma that often confronts anglers who fish with sea worms is what to do with this $3 per dozen bait if not fully used. Rather than attempting a half-baked approach at salvaging the worms only to have them die soon after, start by keeping your worms cool and moist. If you are lucky enough to have access to a refrigerator, keep the worms in a flat, cardboard box used for beer or soda cans. Line the box with newspaper to soak up the excess moisture and cover the worms with seaweed. Moisture and the sun are what will kill worms quickly. Besides a 'fridge, other good dry, dark areas to keep worms cool are basements and garages.

Usually fashioned into strips that are typically drifted alone or jigged with a bucktail, squid is the bait most often used for weakfish by the general populous. For weakfish, most generally

Did You Know?
Weakfish are members of the croaker family. Male weakfish possess the croaker family's ability to produce a drumming or croaking sound.

like to drift strips about 6-inches long but slightly wider than the usual fluke strip, being about ¹/₂-inch wide rather than ¹/₄-inch. For bucktails, keep the strip's length the same but cut it about □ ¹/₄-inch wide so it flows nicely with the bucktail. Hook the strip once near the top so its tail flutters. Double hooking usually results in a spinning or unfluttering bait.

Though most anglers use regular defrosted squid, some anglers baste their squid strips in a briny mixture of kosher salt and water to bleach and firm the strips. Other anglers dye their squid strips with yellow or red food dye. With this trick, they get the benefit of a differently colored bait, which weakfish often find attractive, plus the squid strip's scent and inherent flutter that jelly worms or Fin-S-Fish cannot duplicate.

Additional weakfish baits include: snappers, peanut bunker, small butterfish, short "shoe-lace" eels, finger mullet, small herring, spearing and sand eels. Anglers typically liveline these baitfish via the same livelining rig employed for striped bass, or they drift them with a fish-finder rig. Anglers can catch snappers and eels with light tackle while catching the other baitfish via a scoop net, drop net or cast net around the bay's marsh banks, lagoons, docks, piers, and sand bar ledges.

Tackle & Techniques

The best rod and reel combinations to tackle weakfish with in the bay are 6 to 7-foot, stiff-tipped, 100% graphite rods rated for 10 to 20-pound line. For drifting baits, anglers typically employ small, lightweight, conventional reels designed for freshwater use. Experienced anglers prefer to employ a spinning outfit so they can drift bait or cast bucktails by simply changing tackle rather than having to bring a second rod.

Anglers typically spool their reels with 6 to 12-pound test. Six pound test is used when targeting spikes or when casting lures in shallow clear water locations where weakfish can possibly see thick line while twelve pound test comes in handy when livelining baitfish for tiderunners. Ten pound is the best overall choice because it gives anglers the ability to cast far, because it's relatively light, while having the strength to let the angler successfully lock horns with any weakfish the bay can throw at you.

When drifting bait for weakfish, anglers typically employ either the standard fish-finder rig, that allows maximum sensitivity, or the hi-lo rig, which lets anglers work one bait near the bottom and another higher in the water column. Page 36 discusses tying these two rigs.

The same basic fish-finder rig employed for fluke works great for weaks. Most anglers use the same English wide gap style hook and leader they would for fluke because weakfish have relatively large mouths. while others exchange the wide gap for a beak or Carlisle style hook thats size is dependent upon the size of fish the angler is targeting. A 1/0 is appropo for tackling large weakfish while a #5 is the ticket if the angler is planning on encountering spike weakfish.

The same hook selection applies to the hi-lo rig. With this rig though, experiment with two different type baits to see what the weaks are partial to that day. Fish a squid strip along the bottom, which will appeal to both weakfish and fluke, while using a

shallow stream after their quarry, so why should we approach weakfish, which are perhaps more timid than trout, any differently? Try snelling a #8 wide gap directly to your line with a

more weakfish oriented bait, such as a sandworm, on the hi hook.

Regardless of whatever rig you choose to employ, avoid rigs adorned with beads and spinner blades or tied with heavy, highly visible line. I admit to occasionally stringing one or two pearl beads ahead of my hook when weakfishing, but some of the "contraptions" marketed for weakfish appear more like miniature space probes than things that would attract a weakfish. In fact, these type rigs often frighten weakfish more than they attract them.

When weaks are especially finicky, try fishing for them the way savvy trout anglers stalk trout. One does not see these anglers blazing down a

small split shot placed about 18-inches beyond the hook. Take a whole, live sandworm, insert the hook into its mouth, thread the worm up the hook, and then pop the hook's point out from the worm. Drift the worm at areas where weakfish wait for a passing meal, such as a creek mouth or the edge of a flat or sand bar. This rig's lack of hardware makes it good to employ when weakfish are very line shy and usually keeps crabs and sea robins away.

With weakfishing, hooking the bait is very important. Properly hooking the bait conceals the hook from a weakfish's discerning eyes while also firmly cementing the bait to the hook.

When hooking a shedder crab chunk, simply insert the hook into the

When weakfishing gets into full swing during July, it is possible to stop at many locations throughout Barnegat Bay and catch a few weakfish in a short period of time via lures. A leadhead and worm combination is especially effective for weakfish (center).

chunk's appendage hole and thread the chunk up the hook's shank and pass the hook's barb out the leg socket, being sure to conceal the hook as best you can within the crab.

For hooking sea worms, insert the hook into the worm's mouth and bring the hook back through its head, about a $1/2$-inch down so that the worm is firmly on the hook and will hang off straight.

For squid and spot strip baits, hook the strip once through the strip's squared end and make sure, by gently dragging it slightly below the surface, it runs straight with an occasional flutter and does not twirl.

Chumming

Chumming is a highly productive avenue for catching weakfish. If done correctly near a good weakfish area, chumming with shrimp is arguably the most effective way to catch weakfish. Weakfish chum usually consists of fresh grass or sand shrimp with some anglers combining their shrimp with mashed up blue claw crab, rice or bunker oil. Anglers typically chum during a moving tide so the chum flows with the current and draws weakfish towards the angler's position. Hooks baited with shrimp or shedder crab are then drifted back into the chum slick similar to the way bluefish anglers fish their slicks for bluefish. Once weakfish pick up the chum's scent, anglers can usually enjoy solid fishing.

When chumming from a boat simply travel to a proven weakfish hotspot, like Tices Shoals or Gulf Point. When you arrive, pick an area to anchor so your chum will seep to a likely weakfish holding area, such as a ledge while trying to minimize engine use so you don't frighten any weakfish. Double anchor the boat, to minimize the boat's swing, upcurrent from the selected area by facing the bow into the current and dropping one anchor from the bow and then a second anchor from the stern.

Once anchored, start your chum slick by taking about 10 shrimp and dropping them off the side of the boat. If the shrimp are alive, squeeze them slightly with your hand to stun them because live shrimp tend to swim off. Maintain the slick by dispensing 10 shrimp every minute or two until you start getting bites. Once bites occur, drop down to about 5 shrimp. Basically, this chumming technique is very similar to the one employed by bluefish and tuna anglers. The only difference is you are chumming with shrimp rather than bunker or butterfish.

While the slick is being established, you can keep yourself occupied by attaching a bobber to a line baited with a grass shrimp and let the line drift back into the chum slick. You can then cast around either a Fin-S-Fish or bucktail, while watching the bobber, and use the lure to determine where the fish are located in relation to the boat. Once the bobber goes under or a bucktail gets nailed, it's time to retire the casting rod and fish the chum slick with shrimp.

Fishing the chum slick basically consists of drifting shrimp baited hooks in the slick. Snell a thin, fine wire long shanked hook, such as a #4 or #6 Aberdeen, directly to your line and bait it with two or three large shrimp. Hook the shrimp by inserting the hook underneath the back of the shrimp's tail and exiting the hook out the shrimp's head, being careful to thread the shrimp up the hook's shank so the multiple shrimp will hide the hook. Besides shrimp, anglers also fish shedder crab chunks, especially if their chum contains crab parts. Although it is recommended that you employ no weight, some anglers like to include a small splitshot usually about 10 to 15-inches above the hook to lower their baits to where the fish may be holding. If you find weakfish holding low, try the sinker. Otherwise, stick to avoiding weight because weaks will primarily focus their attention on the chum which your unweighted bait will drift with. Now it's time to fish the slick.

Toss 10 grass shrimp in the water and allow your shrimp-baited hook to drift with the "chum" shrimp by free spooling line. Keep track of how much line you ladle out before you get a hit, so you can zero-in on where the weakfish are holding within the slick. When fishing with shrimp, be prepared for slight, almost undetectable bumps because weakfish will often pick your shrimp up very gently. Unless you are really paying attention, you will not even feel the weakfish until one decides to run with the bait. Stay focused, especially in the beginning of the trip when most anglers get "antsy" if fish do not immediately smack their baits.

Even if you miss a few weakfish, a good chum slick will usually attract such a large number of them that you will have more than enough chances to relax and enjoy the fruits of your labors.

Livelining

Livelining baitfish for weakfish is very similar to livelining baitfish for striped bass with the only differences being the baitfish are typically smaller in size, because weakfish tend to be smaller than bass, and the tackle is lighter. Spot, finger mullet, peanut bunker, small snappers, small herring, and small eels are the usual livelining choices with spot being the perfect choice if you are bent on catching a tiderunner.

Anglers liveline baitfish for weakfish similar to how striped bass anglers liveline their baitfish: employ suitable tackle so one can present the baitfish swimming unimpaired as it would in nature. The bronzed Mustad short-shank tuna hooks, in the 3/0 to 6/0 size, and light, thin-wire hooks, such as a #2 Aberdeen, or small treble hooks are good for livelining baitfish for weaks. Whichever hook you choose to employ, make sure the hook is not too large for the baitfish. You do not want to mortally wound the baitfish so it does the dead man's float near the surface. Many savvy anglers simply snell a short shank tuna hook directly to their main line and avoid hardware of any kind. Others, take a two to three foot piece of 15-pound leader material and tie it to a small, black barrel swivel to prevent their line from becoming twisted with their bait's movement.

Hooking a baitfish for livelining is extremely important because if done incorrectly, you run the risk of either killing it or flipping it off the hook when casting it. Either insert the hook just before the baitfish's dorsal fin, making sure the hook goes no deeper than

Did You Know?

Two anglers share the all tackle record for weakfish at 19-pounds, 2-ounces. One weakfish was taken at Long Island back in 1984, while the second record fish, was taken from the Delaware Bay in 1989.

½-inch, or hook baitfish through both its lips. A savvy angler hooks a spot by inserting a hook into the fish's mouth and then out its cheek because weakfish, according to some, hit their prey head-on.

As stated earlier, when livelining a baitfish, allow the baitfish to swim near structures or locations that weakfish would likely be patrolling, such as a shoal or bridge piling. Keep your reel dis-engaged so the baitfish can swim about unrestrained, and a weakfish can easily take line once it grabs your baitfish. When a weakfish takes your baitfish, let it take line and wait for the fish to pause. Weakfish, unlike most striped bass, will usually drop and re-grab a baitfish a few times before finally attempting to eat it. When the weakfish pauses a second or two, to turn the baitfish around to swallow it head first, engage your reel and wait. When the weakfish moves once again, set the hook.

A Look at Lures

By now you should have noticed a pattern with lure selection. With weakfish, the pattern of matching the size and type of your lure to the area's prevalent baitfish is followed more astutely because weakfish are usually much more finicky than bluefish or striped bass.

Among the lures weakfish anglers cast around Barnegat Bay, Fin-S-Fish, Need-L-Eels, Nordic Eels, Rat-L-Traps, Bomber plugs, and bucktails account for the consistent catches. Use these lures only as the foundation for your lure arsenal because many times you will have to stray from this list to match the baitfish the weakfish are feeding on. For example, when weakfish are focused on eating very small baitfish, such as rainfish, the savvy angler will cast shad darts, 1/8-ounce Bridgeport jigs or kastmasters to fool the weaks. Anglers casting large bucktails or plugs will often get skunked while the anglers casting these small lures catch fish.

When casting lures at a bluefish school attacking baitfish on the surface, look for weakfish to often hang below the blues on the bottom. Weakfish, unlike bluefish, do not typically prowl the surface, causing a large commotion that acts as a beacon for lure casters. Fin-S-Fish, Need-L-Eels, Nordic Eels or bucktails are a few lures one can use to drop below the bluefish chaos to the bottom where the weakfish are likely feeding.

Within recent years, the Fin-S-Fish has become a popular weakfish lure, rivaling bucktails in many scenarios as the lure of choice. A Fin-S-Fish is a soft plastic fish marketed by Lunker City which anglers thread onto a leadhead jig and slowly bounce along the bottom bucktail style. Colors, such as gray sparkles, red sparkles, shad, and green sparkles, in 4 or 5³/₄-inch sizes, are good choices to use because they resemble weakfish forage.

Similar to the Fin-S-Fish, anglers bounce regular 1/8 to 1¼-leadhead jigs with either a 4-inch "freshwater style" twister tailed grub or jelly worm. Once in a while, try forgoing the artificial jazz and bounce a whole sandworm with a plain leadhead jig. Initially you want to securely thread the worm on the jig by inserting the jig's hook into the worm's mouth. You then thread the worm halfway up the hook's shank and then pop the hook out the worm's side. The worm should be fully extended along the hook's shank so the only parts of the jig exposed are the jighead and the hook's bend. Work the jig the same way you would a Fin-S-Fish or bucktail — slowly along the bottom.

with the former two being the more popularly used weakfish lures in the bay. When selecting sizes and colors for these lures, remember to match them to the size and coloration of the baitfish the weaks are feeding on.

For Rat-L-Traps, the standard chrome with either a blue or black back

Need-L-Eels and Nordic Eels are two lures that anglers should have handy if they happen upon a sand eel school because both lures are designed to match sand eels, a weakfish stable. Need-L-Eels are flat, thin lures while nordic eels are metal-headed lures with a split surgical tube tail. Anglers typically retrieve these lures along the bottom with an occasional slight but sharp twitch that makes the lure dart like a sand eel.

Rat-L-Traps, MirrOlures and Bomber plugs round out the lure list

in the ½ to 1¼- ounce sizes generally work well.

Silver and green backed, blue backed and black backed are good color choices for a MirrOlure, Bomber or Rebel aimed at weakfish. Green and blue backed lures are effective when the weaks are feeding on herring, bunker, or rainfish; black backed lures are best used when mullet are the predominant baitfish. For MirrOlures, try a 52M model; for Bombers, try a 15A or 16A model while for Rebels try a 4½-inch F-20 model. Work these lures slow as

"Spike" weakfish abound in Barnegat Bay and are fun to catch on light tackle via lures, such as bucktails and Fin-S-Fish (center).

one would for striped bass. Do not use a swivel or wire leader because these will partly impair the lure's built-in action. Save for a weakfish's two vampire-like fangs, they do not possess a bluefish-like mouth that can sever lines.

Channel edges, rocks, bridge pilings, and the edge of sand or eel grass flats are the prime areas to work these lures for weakfish within the bay. These type lures, which rely upon horizontal movement, are very hard to work when weeds abound. So, if weeds become too thick on the surface, tie on a jig type lure and drop it below the weeds.

Buck Tales

Bucktails are the most universally utilized lure among lure casters and are probably more widely employed for weakfish than for any other species within Barnegat Bay. A bucktails is an effective lure for weakfish because its small, streamline silhouette resembles the outline of the small baitfish weaks savor.

Bucktails come in a wide array of colors and sizes. Pure white, white mixed with red and pure yellow are popular colors for weaks while the lima bean, open-mouthed smilin' bill and cannonball are popular bucktail leadhead shapes. Weakfish typically flip for small 1/8 to 1¼-ounce pure white bucktails jigged slowly along the bottom. Though some anglers go as high as 2-ounces, you will often find sticking to smaller bucktails will often produce better results because weakfish tend to prey upon small baitfish. Using large bucktails on the bottom of a hi-lo rig or as a sinker substitute generally perform better than when jigged alone. As with

other lures aimed at weakfish, savvy anglers do not use swivels when jigging bucktails so as not to hamper the bucktail's movement nor spook a potential catch.

After selecting a bucktail to jig, you then want to select a "sweetener" to add scent and action to the bucktail. Sweeteners also make a weakfish more likely to hold onto the bucktail for a few critical seconds so you can set the hook. Popular bucktail sweeteners for weakfishing include: sea worms (whole and in pieces), squid strips, shedder crab chunks, mackerel strips, Uncle Josh pork rind strips, plastic worms (particularly Mr. Twister), plastic grubs, and spot strips.

When applicable, be sure to hook sweeteners only once and when pickings are slim, place a few drops of either shedder crab or shrimp oil on the sweetener to impart a strong scent to it. Keep strip baits and sea worm pieces in the 4 to 6-inch range. With jelly worms and grubs, use as is but if you find weakfish nipping them in half, shorten their length. Do not, however, shorten them too much because it will often cause weakfish to become disinterested.

Bucktailing is best done with a stiff-tipped, 6 to 7-foot graphite rod rated for 8 to 20 pound line. The rod's stiff tip and graphite composition will allow you to set the bucktail with a quick snap before the weakfish has a chance to spit the bucktail out. A spinning reel with a smooth drag, loaded with 8 to 12-pound test is preferred. Many anglers like to use 10-pound test, which is light enough to cast yet strong enough to bring in most weakfish without any great angling skill.

Work a bucktail for weakfish the same way you would for fluke: slow, short steady vertical jigs so the bucktail pokes along the bottom in small 2 to 4-inch hops. Occasionally alter the speed of the jigs to make the bucktail dart and to determine what action best entices the weakfish. When you feel a weakfish grab your bucktail, which usually occurs when the bucktail falls back to the bottom, quickly set the hook with a quick yank before the weakfish spits the bucktail out. Weakfish, like fluke, will not hold a bucktail too long.

Where and When

Weakfish occur along the Eastern Coast of the United States, from Massachusetts to Florida. Their greater concentrations can be being found between the Chesapeake Bay and New York State's Peconic Bay.

Weakfish prefer the mouth of creeks, rivers, channel edges, sand flat edges, bridge pilings and eel grass flats. One can find weakfish over a variety of bottoms, but they seem to prefer shallow sandy bottom areas near eel grass.

Weakfish are notorious for being extremely sensitive to their environment. They are not entirely a nocturnal species like striped bass, but are generally caught in the off hours because they do not like the surface commotion associated with boat traffic. The hoopla Barnegat Bay experiences on a given Saturday afternoon is enough to send most weakfish to the hills. The best weakfishing usually occurs from daybreak to about 8 am and again at dusk and night. This is not to say fishing for weakfish at high noon is a waste of time; it's a matter of odds. Similar to horse racing, if you fish at daybreak, you're the 2 to 1 horse whereas the angler fishing at high noon, amongst the wakes from passing boats, is the dark horse going off at 40 to 1. Sure the 40 to 1 horse may come in first and the 2 to1 horse may not, but if you had to bet on the probable winner, who would you pick?

Weakfish usually begin to appear in Barnegat Bay around late April. Weakfish catches become more common as May wears on, and June produces steadier catches as weaks spread out into the bay. July is usually a productive month, but the best weakfishing usually occurs from August to mid-September. Especially during September, the bay's waters are still warm and largely undisturbed by boat traffic and the weakfishing is superb. The majority of weakfish will hang around the bay into late September and begin to leave around mid-October. Some will stay until late October and even to early November. After the first week of October, weakfish anglers for the most part switch over to dunking eels for stripers or probing inshore wrecks for tog and leave weakfishing until the next summer.

Upper Bay

Hot weakfishing areas dot Barnegat Bay's expanse with the mouth of the Point Pleasant Canal and Gulf Point being the two bookends that bracket numerous other spots. Anglers roaming Barnegat Bay's north end for weakfish should investigate the Point Pleasant Canal's mouth. Here, anglers catch weaks by drifting bait or jigging

Did You Know?

A 22-inch weakfish is approximately 5 years old.

bucktails and Fin-S-Fish. The #1 marker and the edge of the shallows that lie south of the canal are usually productive areas. Chumming with shrimp and crab parts is often good in the waters around the mouth but wait for the tide, that surges through the canal during a moving tide, to calm down before attempting to chum. You want to chum with moving water but not with a raging tide that will quickly dissipate your chum.

If you find the canal area too crowded with boat traffic or having too swift a tide, you should investigate the Metedeconk River, which flows just west of the canal. Casting bucktails or setting up a shrimp chum slick is best done at the river's mouth where weakfish tend to concentrate. Two popular areas to fish at the Metedeconk are the waters where the river's northern tip borders the Beaver Dam Creek and at the relatively shallow water that flows between the Metedeconk's southern tip and Herring Island. If weakfishing is your game and the Metedeconk is the place, fish very early or very late when the river is quiet. Otherwise, you'll be looking at slim pickings.

If you journey around Herring Island and fish the island's eastern side, you will come upon a channel that extends past the 10-foot mark and flows past the shallows surrounding Herring Island. Drifting sandworms along the channel's edge is a good way to catch the area's weakfish but be prepared to take a few sea robins off the hook. Anglers can also find weakfish prowling the island's shallows during the off hours particularly at high tide. Herring Island is best fished during the week because the channel that winds past it is heavily

traveled, especially on weekends.

South from Herring Island, we come upon the Mantoloking Bridge whose structure hosts a significant weakfish population throughout the summer and fall. Anglers can find weakfish lurking around the bridge primarily at dusk, night and daybreak. Once 8 or 9 o'clock rolls around bringing boat traffic with it, weakfish largely become inconspicuous until dusk. Anglers can and often do, however, catch weakfish during the day at the bridge, but the better fishing is to be had during the off-hours.

When fishing the bridge, anglers work lures, typically bucktails or leadhead/sandworm combinations close to the bridge where weakfish typically lay to ambush their prey. A whole sea worm threaded onto a plain leadhead and slowly jigged along the bottom is a very potent combination when fishing at the Mantoloking Bridge. Some anglers prefer to drift sea worms here, usually via a fish-finder or hi-lo rig, primarily concentrating their efforts at working the edges of the channel before and after it flows beneath the bridge.

Besides being attracted to the bridge's foundations, weakfish also take a liking to the several holes that dot the bayscape south of the bridge. Look roughly a few hundred yards south of the bridge, to find these holes. Locating one shouldn't be that difficult because a few drop past the 12-foot mark while the surrounding water is around 7-feet deep. Anglers with depth recorders can simply employ their machines to locate the holes while anglers without such a device can either drive around and drop bucktails to note the depth or can see where boats begin to congregate. Drifting the holes with hi-lo rigs, to catch both weakfish and fluke or chumming with shrimp are generally productive if you stick to the early morning. Once the sun rises and boaters start raising a ruckus, things usually go down hill quickly.

South from the Mantoloking Bridge lie a trio of locations worth careful investigation: the Route 37 Bridge, Pelican Island and the Toms River. Similar to the Mantoloking Bridge, the Route 37 Bridge's foundation plays host to a variety of baitfish, and weakfish are one of the many predators that stake the bridge out for meals. Experienced anglers work the channel's edges as it flows beneath bridge, and the bridge's structure. When fishing the Route 37 Bridge, anglers should explore the channel that runs parallel to the bridge's northern side. This channel, which eventually flows past Pelican Island is usually productive for weakfish for anglers using squid strips or sea worms.

If crabs become a nuisance, which they sometimes become here, switch over to lures or attach a small float to your bait. Otherwise, you might as well settle down and try crabbing. Whether or not to employ a small cork float a foot before your bait to avoid crabs is a tricky call. I would tend to say no to using a float when weakfishing because it seems the bay's weaks become wary of baits rigged with floats. However, there have been times when anglers fishing with floats have cleaned up while others spent the day fighting off blue claws and sea robins.

Pelican Island, which the Route 37 Bridge spans to, is generally productive for weakfish if you target the waters north and south of the island's bridge and the waters off its southwestern shores. There is some deep water off the island's southwestern side that lies close to the island and is good to fish early in the morning. Though spike weakfish abound here, patient anglers employing live or cut bait will occasionally hook into a 4-pound fish. Early or late in the day are usually the better times to target this location. The one problem with Herring Island, just like the Route 37 Bridge, is it tends to get crowded rather quickly with boats on the water and weakfish usually retreat to the flats or find some other peaceful hiding place. As stated earlier, don't throw in the towel just because it's mid-day and boats are flying by. Many times, I have experienced great weakfishing when the sun is high and the boats are chewing up the surrounding water.

Southwest from Pelican Island flows the Toms River, which attracts a sizable weakfish population. Though the majority of the river's weakfish are of spike stature, anglers working the river's mouth will typically score with larger, legal-sized fish. Usually, the top areas to target in the Toms River are Coates Point, Long Point and Goodluck Point. Many anglers find fishing the Toms River's mouth frustrating because the river seems to keep a relatively steady depth and does not possess an obvious weakfish magnet, like a Meyers Hole-like depression. At the mouth, you have a stretch of bay that contains slight bottom contours which weakfish often hold near. Though these contours are not as apparent as other weakfish attracting locations, anglers who

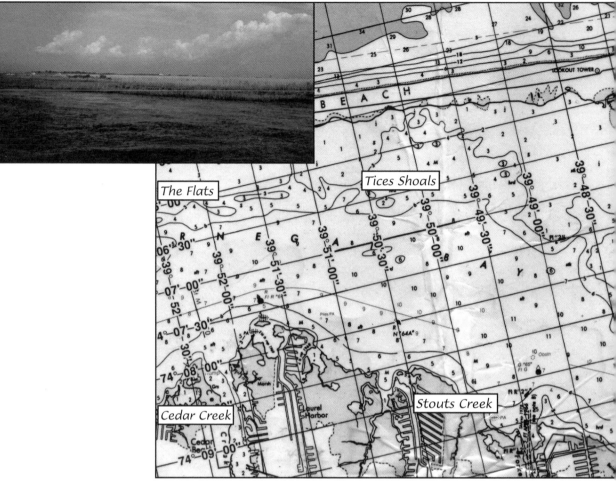

The Flats

Tices Shoals

Cedar Creek

Stouts Creek

Experienced weakfish anglers usually concentrate their efforts at Reeves Cove, the flats, Tices Shoals, the mouth of Cedar Creek, and Stouts Creek (center). When poking around the bay's eastern flank, look for weakfish to hold downtide from salt marshes and creek mouths (top left).

explore the waters near the "Points" will find weakfish holding contours and slight depressions.

Southeast from the Toms River lie the flats that guard Island Beach's western shores. These should be the focal point of any weakfish angler plying Barnegat Bay. Starting roughly at Seaside Park and ending at Oyster Creek Channel, the flats attract weakfish because they support a diversity of forage and are largely undisturbed by humankind. One suggestion when fishing the flats: cut your engines to a slow, no-wake pace or ideally drift so as not to scare the many weakfish holding

at the flats' perimeter. Ironically, the perimeter is where most boats are seen blaring their engines. By driving very slowly up to a spot, you will not spook the weaks as easily.

From an angling perspective, Reeve's Cove, forming the flats' northern boundary, is the first area that attracts attention. Reeve's Cove is a T-shaped depressions that extends east into the flats. Considering the cove's depth is 5 to 7-feet while the flats to the north, east and south have a 1 to 3-foot depth, it's easy to see why anglers who fish the cove catch weakfish. At the cove, weakfish can usually be found looking

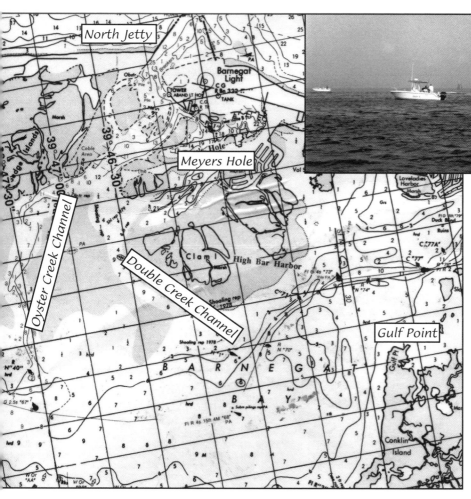

Labels on map: North Jetty, Meyers Hole, Oyster Creek Channel, Double Creek Channel, Gulf Point

for prey near the ledges, where the cove's deep water rises to the flats, or near the small contours or holes located on the flat itself. At high tide, anglers with small, flat-bottomed boats often venture onto the flats and look for the small holes and contours which typically hold weakfish.

South of Reeve's Cove, anglers should try for weakfish along the flats' entire edge. If conditions permit, drift towards the flat and avoid using your engine so the area's status quo is undisturbed.

Following the flats south, one will eventually arrive at Tices Shoals which is one location serious weakfish anglers should investigate. Though the flats extend further south to Oyster Creek Channel, Tices Shoals constitutes the flats' southern boundary to many anglers. Shrimp, spearing and blue claw crabs teem within the waters around Tices Shoals, and weakfish maintain a strong presence at the shoals making it a great place to fish.

The key to effectively fishing Tices Shoals is to fish the slopes where the water rises from 6 to 2-feet deep. Especially during an outgoing tide,

When seeking weakfish in the lower bay, anglers should try the Mud, Oyster Creek and Double Creek Channels; Barnegat Inlet and Gulf Point (center). Leave early for these locations, particularly the channels, because they often become clogged with boats very early (top right).

anglers can usually find weakfish hanging downtide from the slope. Anglers typically anchor their boats in the deep water so their chum or baits drift along the slopes. The rising tide is also good to fish because it allows anglers captaining large vessels to maneuver more easily, and the weaks seem to become more active.

At the shoals, chummers generally have a productive time when employing shrimp chum and shrimp or shedder baits. Usually during an incoming tide, chummers anchor their boats in deep water and let the current bring their chum onto the flats. Lure casters should try casting up onto the flat, letting the lure settle and then doing a slow retrieve or bounce back towards their position bringing the lure past the ledge.

Across the bay from Tices Shoals lies Cedar Creek which is a very popular weakfish location with the better areas to fish located at the creek's mouth. For landbound anglers, the northern tip of Cedar Creek's mouth, where Berkley Island County Park lies, is a good area to find weakfish and net weakfish baits. For boaters, the southern end of Cedar Creek's mouth, off Cedar Creek Point, is a good place to cast lures or drift at daybreak. Fishing the creek is generally best during an outgoing tide at night or daybreak. The tide, coupled with the creek's natural flow, forces baitfish out of the creek and into the bay where weakfish and hopefully you are waiting. At night, weakfish creep very close to the creek, occasionally being found farther west into the creek than thought possible.

A little south of Cedar Creek flows a smaller creek named Stouts Creek. Though Stouts Creek's mouth is not as wide or deep as Cedar Creek's, it still manages to attract weakfish that wait there during an outgoing tide to ambush the creek's baitfish. Boaters should try a drift or two where the creek extends into the bay and drops in depth near the Intracoastal Waterway. There, and where the creek's shallow water drops down to the bay's deeper water is where you will generally find weakfish.

Lower Bay

When experienced anglers consider the lower bay for weakfish, places like "Meyers Hole" and "Gulf Point" frequently get mentioned. Year after year, these two locations consistently provide great weakfishing for countless anglers. What makes the lower bay so special is that it contains these two prominent spots plus other areas that provide great weakfishing.

One location that does not glare out as a weakfish spot but has a small following of anglers is the Forked River. Similar to other creeks and rivers mentioned so far, the Forked River's mouth is where the weakfishing action is to be had. Anglers drift squid strips or cast bucktails off the river's mouth, at dusk or very early in the morning when the boat traffic is light, are the ones who consistently score with weakfish.

To the south of the Forked River flows Oyster Creek which is a popular location for weakfish. The Oyster Creek Power Plant is a very large landmark that makes it easy for anglers unfamiliar with the bay to locate both Oyster Creek and Oyster Creek Channel. At dusk or daybreak during an incoming tide, anglers can often catch weakfish as far west as the Route 9 Bridge, which spans Oyster Creek. Most weakfish though, are caught where the creek empties into the bay. Oyster Creek neophytes should investigate the waters between the creek's mouth and the #67 buoy, which marks the entrance to Oyster Creek Channel.

As its name implies, Oyster Creek Channel flows close to Oyster Creek. Starting at the channel's western mouth at buoy #67, anglers primarily drift with sea worms and squid strips or cast lures for the weakfish that prowl the waters off Oyster Creek. This area is good to fish at dusk or early in the morning before boats starting churning up the water. One may not want to take their act any further into Oyster Creek Channel.

Don't overlook the possibilities within the channel itself. Weakfish will often hold near the bottom contours and even near the blowfish coral beds that lie within the channel. See Chapter 7 for information regarding blowfish coral. The channel's edge, where it rises up to the shallow eel grass flats, is very productive for weakfish.

Oyster Creek Channel's eastern mouth, where it flows parallel to Island Beach before reaching Barnegat Inlet, is a good location to target with strip baits, bucktails, Rat-L-Traps, or livelined baitfish. The few creeks that flow from Island Beach into the channel are good places to look for weakfish. Savvy anglers liveline baitfish that appeal to weakfish, such as spot or small herring at the pockets off the creeks where striped bass anglers liveline eels.

A tad north from Oyster Creek lies the Mud Channel. What makes this relatively small channel so special for weakfish is its close location to the shallows behind Island Beach where weakfish abound. Approach the Mud Channel as you would the flats: avoid using your engine and work the channel's edges. Many times weaks hang dead center in the channel so don't rely solely on the channel's edges for your catch. Chances are good that you are floating right above them. The best time to fish the Mud Channel is at dusk or daybreak and when totally devoid of boats. One loud, errant boat that blows through the channel will definitely throw a monkey wrench in the works.

Oyster Creek Channel eventually empties into Barnegat Inlet. The inlet holds many possibilities for weakfish anglers with Meyers Hole being the most popular weakfish spot in the inlet and perhaps in the entire bay.

The inlet's main channel may not attract as much publicity as Meyers Hole but it can yield large weakfish to anglers willing to devote some time fishing. The better parts of this channel to fish are where it joins Oyster Creek Channel at the inlet's western mouth. Also, the edges of the channel as it snakes around the large sand bar are good to work with lures but patience is key because weakfish will often be there but won't strike until the right bait or lure goes past their nose. Early in the morning or at dusk, when boat traffic is non existent, is the best time to attack the channel.

If you follow the inlet's main channel south towards Long Beach Island rather than north towards the ocean, you will venture past a small marsh island, which is good to fish with lures.

Southwest from the marsh

island lies Meyers Hole which is arguably the most popular weakfish location within the entire bay. During the summer, the number of boats chumming at daybreak or especially at dusk will clearly point out the approximate location of the hole.

Chumming with shrimp is the most popular method of fishing for weakfish at Meyers Hole because the hole is largely shielded from wind and boat traffic by the shallows to its north, land to its west, and a residential no-wake waterway to its south and east. Meyers Hole drops past the 25-foot mark and acts almost like a large collecting bin that holds baitfish dragged by the inlet's strong currents. Weakfish merely hang in Meyers Hole and wait for the tide to feed them. During an incoming tide, anglers position their boats so their chum seeps towards the channel and the sand bars. Casting small plugs, Rat-L-Traps, bucktails, and Fin-S-Fish here is generally productive if you can get past the snappers.

Northeast from Meyers Hole lies the inlet's north and south jetty where weakfishing is usually best at night. Especially during late summer nights, savvy anglers will liveline small baitfish, particularly spot, for the large weakfish that often prowl the jetties. When fishing the jetties, look for weakfish at many of the same areas that hold striped bass. Also, work the area below the bluefish schools that usually raise a ruckus here because large weakfish will hang below the bluefish.

Anglers who want to stick to the bay, should investigate Double Creek Channel, which exits the inlet and flows southwest towards Barnegat Township.

Timing is critical when fishing Double Creek because it's a relatively narrow channel that quickly becomes unenjoyable to fish once boats start tearing through it en route to the inlet. Plan your trips early or late.

Anglers should initially investigate Double Creek's eastern mouth, where it flows past Long Beach Island's marshy banks. Here, the waters are 20-feet deep a scant 2-feet from terra firma. Try fishing this stretch with a hi-lo rig or a fish-finder rig because weakfish and fluke often share the same underwater depressions here.

As you follow Double Creek west towards the bay's western shores, anglers usually succeed when fishing along either the edge of the sand bar that flanks the channel's northern side or the eel grass flats that lie to the south. Along the remainder of Double Creek, as it cuts past Clam Island and heads towards the #68 marker off Barnegat Township, anglers fish the channel's bends and deep pockets located in the channel's interior.

The water between Double Creek's western mouth and the #68 marker are productive for weakfish fishing and usually besieged with boats during the summer. Though the cardinal rule with weakfishing is to fish very early or late to avoid boat traffic, I have caught large weakfish here at high noon with boats blowing through the area, so don't let a few boats or the time of day stop you from trying a few casts or a drift with squid strips at this productive area.

Many of the bay's better weakfishing areas are places that get passed by in the mad rush to the channels and inlet. Gulf Point, located at the northeastern tip of Conklin Island, has a well deserved reputation for providing superb weakfishing but does not attract as many boats as other locations. Playing host to a wide variety of weakfish prey – blue claw crabs, grass shrimp, snappers, spearing and spot, it's not surprising why Gulf Point attracts so many weakfish. With its convenient location near Double Creek Channel and the Barnegat Township Public Docks, Gulf Point is one of my favorite areas to fish for weakfish.

One should plan their weakfishing trips to Gulf Point either during the early morning or at dusk. Because the Point is a popular area to clam and crab, it tends to become crowded during the day especially if the day happens to be Saturday or Sunday. Though you can catch weakfish if there is boat traffic, the odds of a good trip are more in your favor if the water is quiet.

Swarms of boats are not the only swarms you should be wary of when fishing Gulf Point. The resident green fly, gnat and mosquito populations will often attack boaters with the force of a biblical plague. Especially during July and August, make sure to cover up with either clothing, insect repellent or Avon Skin-So-Soft, or you will be in for an irritating experience.

The waters off Gulf Point run roughly 7 to 8-feet deep with the bottom gradually rising to around 5-feet deep as you venture close to Conklin Island. The bottom plateaus at the 5-foot depth and there are several areas of the 5-foot plateau that drop a foot or two forming depressions. These depressions, no matter how small, often hold weakfish. Most depression are small, so

it's hard to locate them by driving around without scaring the fish. To avoid frightening weakfish, many anglers opt to drift the Point to feel the lay of the land.

Many anglers have luck at the Point's 7-foot deep areas especially during the day and when the bay's water temperature heats up in late summer. The patches of eel grass that float around Gulf Point may give lure casters gray hairs but warrant scrutiny because the grass provides a great habitat for weakfish prey whether it's floating in large clumps along the surface or resting near the bottom.

While at Gulf Point, anglers should investigate the channel that flows past the Point and winds east towards Long Beach Island. Casting lures along the shallows that border this channel or drifting in the channel itself is usually productive for weakfish.

By exploring Gulf Point, the flats and other weakfish holding areas that dot the bay and employing the baits weakfish crave, you too will enjoy the sport Barnegat Bay weakfishing can provide.

Chapter 5
Tog & Sea Bass

For anglers who relish bottom fishing, tautog and sea bass are the dynamic duo. Both fish share similar habitats, aggressively attack baited hooks, fight tenaciously when hooked, and make superb table fare. With a little investigatory work, anglers can find these two species sharing the same water with many other species that swim in Barnegat Bay.

Tautog

Tautog are a hard-fighting, good-tasting fish that have gained considerable popularity among New Jersey recreational anglers within recent years. To many anglers, tautog are the northeast's version of tropical reef fish. Similar to the tropical reef fish that live amongst coral reefs, tautog haunt natural and man-made structures. While reef fish come in a dazzling array of bright colors, that match the reef's brightly colored coral, a tautog matches the dark browns, blacks and grays one sees when looking at the rockpiles, mussel-encrusted pilings and rusted ship wrecks they favor.

Also known as "tog," "white-chin," "bubba," "oyster fish," "slippery bass," "black porgy," "black moll" and more commonly "tog," Tautog (Tautoga onitis) are a stout, powerfully built fish. They possess large, thick, rubbery lips that cover their large, chisel-like teeth. Tog use their teeth with great effect to seize and easily crush crabs and their other shell-clad prey. Tog also possess thick pectoral fins and a wide tail which allows them to quickly dart into structure when hooked.

As stated earlier, tautog are a dark colored fish, ranging in color from a dark slate to a light steel-gray. Mature males are usually lighter in color than females and possess solid white on their underbelly from tail to chin, hence the name "white-chin." Mature females are usually darker in color and typically possess dark bands that form patterns along their bodies.

Though tautog are believed to attain a weight of 30-pounds, most average 2 to 4-pounds with anglers considering an 8-pounder to be a "bulldog." Within Barnegat Bay, tautog

A "Barnegat bulldog" that fell to a green crab. Note the tog's white chin and underbelly, unique to the male gender (opposite).

range in size from small 1-pound specimens, occasionally caught from the Route 9 Bridge, to 12-pound plus bulldogs, breaking tackle at, Barnegat Inlet's rock piles.

A Blackfish Buffet

A tautog's diet centers around crabs and other shell-clad marine organisms. Crabs make the most potent bait for tog with some crab species preferred over others. Tog will almost always savagely attack hermit crabs but will show a marked preference for other type crabs, such as blue claws, calico, green and fiddler. Aside from crabs, tog will also respond to sand fleas, conch strips, clam strips, oysters, scallops, and sea worms.

Hermit crabs are unrivaled as a tautog bait because tog simply go bananas when presented with a hermit crab. The hermit crabs residing within Barnegat Bay range in size from small snail-like specimens occupying $1/2$-inch snail shells, to large Florida tree crab-looking specimens that occupy 4-inch conch shells. Though their "borrowed" shells are usually a dull white covered with algae, hermit crabs themselves are quite attractively colored usually being a light cream molted with dark orange and sometimes with light streaks of gold.

A hermit crab's relatively small size makes them an ideal bait when "nibblers," such as small tog, are on the scene because unlike other tog baits, that require the angler to wait for the tog to pass the bait back to its crusher teeth before setting the hook, hermit crabs allow the angler to yank on the first nip and usually hook the fish. The one problem with hermit crabs is you better plan on being either quick with your strikes or, if you don't bring a good bucketful, on a quick excursion when fishing one or two hermit crabs at a time.

The responsibility of catching hermit crabs usually falls on the angler because tackle stores rarely if ever carry hermit crabs. Anglers can find hermit crabs near the bay's sand bars and other shallow locations. From a distance, they often appear as snails but upon closer inspection one will see the crab's legs and claws flattened against the inside of its shell. Once caught, hermit crabs keep well in a 5-gallon bucket filled with fresh saltwater. Hermit crabs, unlike other type crabs, do not do too well when out of water. Keep them shaded and change their water every so often, depending on your bucket's crab population to maintain a "breathable" oxygen level.

Besides hermit crabs, calico crabs, also known as lady or leopard crabs, are perhaps the most conspicuous type crab within the inlet. Their prevalence at the inlet makes them a steady tog prey, and anglers employing calicos for bait usually achieve excellent results.

Small calicos with a top shell of about 1-inch can be livelined whole. Simply insert your hook through one of the crab's leg joints and run the hook out the next leg joint. The trick is to keep the crab alive because its movements are what attract a tog's attention. Anglers typically cut 2-inch and larger calicos into chunks green crab style.

Though tackle stores do not sell calicos, anglers can easily catch some if they venture down to either the inlet's north or south jetty. From either jetty's rocks, scrape calicos off the rocks or net them as the inlet's strong current tosses the crabs about. Anglers bent on catching calicos should explore the surf jetties or under rocks, concrete or wood at tidewater locations. Some anglers set traps baited with surf clams for green crabs near rock piles, piers or especially by clam beds where the green crabs love to prey on clams.

and similar sandy bottomed locations where calicos prefer to dwell.

Green crabs, the bait synonymous with togfishing, look similar to mud crabs but are not as large, averaging around 3-inches wide. In Europe, these muddy green to bright emerald crustaceans are sold for human consumption. Anglers use green crabs the same way they do calico crabs: small crabs are livelined whole while larger crabs are cut into halves or thirds.

Around Barnegat Bay, you can purchase green crabs at many area tackle stores or you can find them near

Once caught or purchased, green crabs subsist in a cool, dark, well-ventilated place such as a garage or basement. As with other type crabs, space them out. Congestion and heat are primarily responsible for green crab fatalities. You can also keep green crabs in a large floating killie car, provided you keep them well fed with clams, squid or fish bodies. Keeping them well nourished will prevent them from turning cannibalistic.

Though regarded more as a summertime delicacy than bait, blue claws will also take their share of tog.

A weary angler after wrestling a nice size tautog from its rocky den. Note the tog's erect dorsal fin. By firmly grasping the tog by its tail, one can prevent injury often inflicted by the sharp spines found along the dorsal fin (center).

Remember, you can only employ legal sized crabs for bait. Because the average legal size blue claw is larger than the largest calico or green crab, anglers will initially cut a blue claw the same way as a calico or green crab but will then cut the crab into quarters and then cut the quarters in half so they get eight baits from one crab rather than four. The small chunks are easier to use when small tog abound and bulldogs will grab small chunks just as well. Cutting up a legal size blue claw may be a little tricky compared to cutting up the comparably mellow and much smaller green crab, but a nice blue claw chunk has been known to catch the discerning eye of more than one bulldog tog.

Fiddler crabs, so named for the extremely large claw that male crabs sport, are right below green crabs as the most popularly used tog bait. Contrary to whatever you have heard about fiddler crabs, the male's large claw can inflict some pain. It is hardly the "paper tiger" some anglers make it out to be. While it will not draw blood, it can inflict a nasty pinch.

The two varieties of fiddler crab that find use around Barnegat Bay are the china-back fiddlers, which are the pale cream colored variety sold at tackle stores, and the fiddlers native to the Barnegat Bay area. These dark chocolate brown native crabs are much larger on average than the china-backs, noticeably more aggressive, and survive longer on the hook. One can find these characters by poking around the bay's salt marshes.

Fiddler crabs, similar to hermit crabs, offer anglers the luxury of striking on the first hit and usually hooking the tog. Small tog and bergalls, a small fish that makes a living eating tog baits, will usually nip at fiddlers and are trickier to hook with the quick snap technique that's perfect with hermit

Fashioning a Large Crab into Chunks

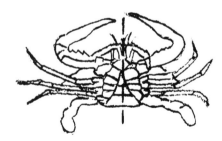 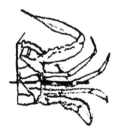 

To prepare a large calico, green or blue claw crab for bait, simply flip the crab onto its back, take a long, sharp knife and cut the crab in half (left).

Cut the two halves in half (center) and use shears to trim off all appendages from the body with a quick snip (right). Some anglers leave the legs intact to make the bait appear more "crab-like." Remove the top shell from the 4 chunks, and you have a tantalizing bait.

Note that the crab chunks should be around an inch long. You may have to only cut small crabs in half to get the proper size bait while king-sized crabs will require much more slicing and dicing. Anglers should save left over shells, claws and legs and employ these "crab parts" as chum. Throwing "crab parts" in the area where you're fishing often produces great results.

crabs. When larger tog solidly hit a fiddler, one quick snap will generally hook the fish.

After a trip, save unused fiddlers by keeping them in a bucket filled a quarter of the way with moist sand topped with fresh seaweed. When the sand becomes dry, moisten it slightly with fresh saltwater. Keep the crabs in a

One can find sand fleas by digging in the sand near the surf at areas like Island Beach or Barnegat Light, which are near the north and south jetties respectively. Simply venture down to where the water breaks at the surf and dig down a foot or so in the moist sand. When you feel a hardy, compact, wiggling creature, you've most

cool, shady place, and they should stay alive for a few weeks.

Sand fleas, also known as sand crabs, are a popular tog bait when employed at sandy bottomed locations that sand crabs and tog share, such as jetties. Sand fleas are typically light gray and average around 3-inches long. Sand fleas hold an advantage over hermit crabs because they can withstand more punishment from a blackfish's chisel-like teeth, but they also require the angler to be more selective of the bites that warrant a response.

likely found a sand flea. To keep sand fleas alive and well, place them in a 5-gallon bucket filled a quarter of the way with moist sand. Keep the sand moist by sprinkling saltwater on it when it appears dry. Do not fill the bucket completely with water because the sand fleas will use all the oxygen and quickly drown.

Besides a wide variety of crabs, tog respond favorably to a few other baits. Small 1 or 2-inch strips cut from a surf clam or conch occasionally grace a tog angler's hook. When bergalls and

The author's brother with a nice November tog. Note this tog's dark, banded coloration which often denotes females (center).

other nuisances abound, anglers simply use the conch's tough, rubber-like flesh to deal with bergalls and small tog.

Oysters, shucked and threaded on a hook, are an unusual tog bait that often find use among veteran anglers who remember that tog are also known as "oyster fish."

rod 6 to 7½-feet long, rated for 20 to 60-pound line is a good rod for tog fishing. The rod's stiff tip lets the angler firmly set the hook in a tog's tough mouth and then maneuver the tog away from any structure. The first moments after hook-up are critical because the tog will attempt to shoot into some sort

Tog will also hit sea worms, but the best tog baits start and end with crabs. When acquiring tog baits yourself, be sure to respect private property, wildlife sanctuaries and any local laws.

A tautog's powerful, stocky build and its love for underwater structure makes fishing for them tricky. Once hooked, tautog will dart into the nearest structure and quickly entangle the angler's line with a quick, powerful surge (center).

Tacklin' Barnegat's Bulldogs

When you couple the "tackle graveyards" tog inhabit with the tog's own tenacity, it's no surprise that heavy tackle is required for wrestling a tog from its den.

A stout, stiff-tipped graphite

of structure, and if the rod is too flexible, it will give a hooked tog enough extra line to entangle your line within the structure.

A conventional reel, with a high gear ratio such as 6:1, spooled with line in the 20 to 30-pound test range will significantly help matters when trying to catch tog. Every foot of line is valuable when tog fishing, and a fast reel lets the angler quickly bring in line that might otherwise be tangled by the tog. The reel should sport strong line that can wrestle a tog away from its tackle-unfriendly habitat while resisting frays. You do not want to fish with very light line, around 10-pound test, because a large tog can probably break this pound test outright without the added luxury of having it nicked from a rock. On the other hand, don't spool your reel with piano wire-like line because heavier line demands more weight to keep your rig stationary, and a heavier sinker increases your odds for snags. The reel's drag should be set with care and its line periodically inspected because a frayed line, coupled with a tight drag, will account for most lost fish that are of considerable size. In any event, a tog's haunt will do such a job on your line that you do not need microscopic analysis to see the obvious frayment in the line. With neglect, the mere action of reacting to a bite will be enough to snap your line. You should also inspect your rod's guides for burs or rough edges because a small burr on the inside of a guide can also chafe your line.

A superb rig to employ for tog in the Barnegat Bay arena, consists of a Mustad Virginia style hook snelled to a 10-inch piece of 30-pound mono leader which is attached to the main line via a drooper loop tied 1 or 2-inches above the sinker. The sinker, usually a bank sinker between 3 and 6-ounces, is connected to the main line with a surgeon's end loop. Some anglers tie a second drooper loop 2 or 4-inches above the bottom loop and then tie a 5-inch leaderd hook for sea bass.

For hook selection, Mustad's Virginia style hook is designed with tautog in mind. It possesses a hollow point and an offset barb that allows for easy penetration of a tog's rubbery lips and hard jaw. This hook also boasts an extra strong shank that resists becoming straightened out when confronted with the pressure of an angler trying to horse in a strong, determined fish.

When fishing in Barnegat Bay, hooks range in size from a #6, intended primarily for small tog, sea bass, porgies, and other similarly small-mouthed fish, to a #3 for wrestling a bulldog tog out from its craggy den. You may be inclined to cheat a little bit with hook sizes, substituting a small size hook when you should really be using one that is much larger. When targeting bulldog blacks, for example, it's better in the long run to use a size #3 because a feisty bulldog can straighten out a size #4 like a pin! Always bring a few larger size hooks, size #3 or #4, because the smaller #5 or #6 size hooks won't usually handle larger tautog. Locations such as the inlet's south jetty often hoodwink anglers with their abundant small tog, and everyone starts tying on #6 hooks only to strike on an "apparent" small tog and lock horns with a bulldog whose brute strength causes it to break free from the undersized hook.

One for the Hooks

To fish hermit crabs, you want to remove the crab from its shell and thread it onto your hook through the worm-like part of its body. To remove the crab from its shell, you don't need to employ some caveman brute force method, such as breaking the crab's shell with a hammer. Initially, try placing the hermit crab in a bucket of fresh saltwater because once in water, the crab will usually protrude from its shell. When out of water, hermit crabs tend to stay squirreled up in their shells making removal, without tearing the crab, difficult. Once you see the crab jut out from its shell, grab it firmly by its legs and claws. Gently pull the crab from its shell by applying steady pressure. If the crab pulls, keep applying steady pressure but don't force the crab because you will likely tear it.

flesh and innards while the chunk's harder leg socket side is positioned opposite the eye.

When fishing with fiddler crabs, one can hook a fiddler a few ways. One way is to insert the hook into the crab's back leg socket and then run

When tog fishing, carefully hook all baits because tog are adept at quickly biting off poorly hooked baits. Note the fish's thick lips which cause dull hooks to not fully penetrate (center).

Maintain steady pressure and when the crab relaxes, the steady pressure usually pops the crab from its shell. Removing a hermit crab from its shell is very similar to pulling a nightcrawler from its hole. Maintain steady pressure but avoid pulling when the worm is pulling downward. Once the worm relaxes, the steady pressure will generally pop it from its hole.

To hook a crab chunk, insert your hook into the natural hole created by the chunk's first leg socket and push the crab chunk up the hook's shank. Position the hook's point in the chunk's

the hook out the crab's front claw. Others simply impale the crab dead center on the hook making sure that only the hook's barb pokes out from the crab's top shell so the crab stays anchored on the barb and does not fall down the hook's gap. When using male crabs, keep the large "fiddle claw" intact because it seems tog often key onto the crab's large claw.

When hooking a sand flea, insert the hook into the flea's head, thread it through the flea's body and push it out the flea's end. Make sure the hook's point slightly protrudes from the flea's underbelly. Otherwise, you will be unable to set the hook properly because the hook will have to penetrate the sand flea's tough exoskeleton before it can penetrate a tog's jaw.

Where and When

Tautog occur from Nova Scotia, Canada to South Carolina with their numbers concentrated between Cape Cod, Massachusetts and the Delaware Bay. Tautog are structure-oriented fish, preferring to inhabit the waters around bridge pilings, rocks piles, wreck sites, bulkheadings, and occasionally clam beds.

Preferring cool water in the high 40s and low 50s, tautog are very conspicuous and aggressive during the late spring and entire fall. In the Barnegat Bay area, tautog first appear between late March and early April at the inshore wreck sites that dot the waters off Barnegat Inlet. As the spring progresses, tautog filter into Barnegat Inlet making the inlet a good place to fish by early to mid-June. The summer months, regarded by many tog anglers as the off season, are very productive if you employ crab baits and stick to cool, deepwater locations in the bay, particularly the inlet. Tog fishing picks up during September and October as the water temperature drops, and the tog prepare to move offshore. November is usually a really good month to fish for tog at the inshore wrecks that rest off of Long Beach Island and Island beach.

Barnegat Bay's Blackfish

At Barnegat Bay's northern end, the Point Pleasant Canal is a good place to fish for tog. Anglers fishing along the canal's sides and near the bridge that spans the canal usually do well with tog. Try fishing the canal during a slack tide because a moving tide in the canal makes keeping a crab bait stationary near structure difficult, even if you are employing a 6-ounce sinker. If we stick to the bay, however, the bridge that spans Beaver Dam Creek supposedly has tog that sulk around its pilings. Though my tog expeditions have never taken me to this particular location, from what I have heard, this bridge will yield decent size tog.

Anglers searching for tog in the upper bay do not need to look too far from the Point Pleasant Canal because the Mantoloking Bridge holds a healthy tautog population. Barnacles, crabs and assorted mollusks cling to the bridge's structure and tog merely peruse the area nipping these organisms off the bridge. At the bridge, tog average around 2-pounds with 5-pounders being an occasional occurrence. When fishing the bridge, anglers anchor up tide from the bridge and "fall back" towards the bridge by letting out anchor line until they are within pole's length of the bridge's structure. It is a good idea to fish the bridge early, before boat traffic erupts, because excessive boat wake pushes the boat about, making the somewhat easy job of anchoring near the bridge difficult.

South from the Mantoloking Bridge, the next tog haven is the Route 37 Bridge. Anglers fishing the Route 37 Bridge should check the bridge's pilings near where the channel winds beneath the bridge. Similar to the way you would fish the Mantoloking Bridge, fish close to the Route 37 Bridge's pilings because

LORAN TD's for a few inshore wrecks:

Caddo:
26891.4/43207.5

Cedar Creek:
26912.8/43339.2

Chaparra:
26847.6/43239.9

Great Isaac:
26840.9/43195.2

Peresphone
(Northeaster):
26897.1/43287.6

Schooner:
26889.4/43239.8

that's where the tog will likely be holding. Fishing at the Route 37 Bridge may be tricky during the day because boat wake will knock the boat around when you are trying to anchor near the bridge. Remember not to anchor in the channel; it's illegal.

South from the Route 37 Bridge, anglers who fish the Route 9 Bridge occasionally report catching tog, usually during the late summer. In my opinion, the bay's best tog fishing is to be had from the rock piles in Barnegat Inlet.

Though anglers can find a few rogue tog near the sedge banks that line Oyster Creek, the prime locations to catch tog are at the rock piles that form the inlet's north and south jetty. Fishing the areas within the inlet is best done during a slack tide when the current is weak. You do not need too much weight to hold bottom and there are not as many weeds in the water. During a moving tide, the current is so strong that your line gets pushed around unless you start using 6-ounce sinkers.

Perhaps the best area to fish in the inlet is the rock pile that runs from Andy's Tackle Store, rings the Barnegat Lighthouse and then merges into the south jetty. Arguably the best togfishing available in southern New Jersey is off the small jetty that connects to this rock pile. The bottom near the base of the jetty is about 25-feet deep and then slopes down to around 40-feet out by the jetty's tip. There are three good holes that lie off this jetty. The most proficient way to locate these holes is to cast around the vicinity of the jetty about 40 to 50-feet out and once you feel your sinker go down deep, you have located a hole. When fishing the jetty's holes keep your bait on the bottom and absolutely still. Keeping it still allows you to feel large tog grab the bait and "slide the sinker." Meaning, they take the bait firmly in their mouth and move off with it, without attempting to nip it off the hook.

Across the inlet lies the North Jetty which is a another good tog area, though it's not usually as potent as its southern counterpart. Anglers typically fish the north jetty from its rocks by anchoring their boat on Island Beach's sandy inlet beach and walking to the jetty. Alternatively, anglers can exit the inlet and anchor off the jetty's northern side. The jetty's northern side offers anglers calm water to easily anchor and fall back to the jetty similar to the way striper or bluefish anglers anchor for livelining or chumming.

If the weather conditions cooperate and you command a somewhat large vessel equipped with a LORAN unit, there are numerous inshore wrecks, mentioned on the previous page, that dot the waters within 3 miles of Barnegat Inlet. While these inshore wreck sites mark the final resting place of fallen sea farers, their mussel-encrusted remnants also play host to a diversity of fish, especially tog.

Black Sea Bass

Black sea bass are beautifully colored fish that haunt the same tackle graveyards as tautog. Also known as "rockfish," "rock bass," "black perch," "talywag," "black will," and more commonly "sea bass," Black sea bass

(Centropristis striata) are one of the bay's better tasting fish.

Sea bass resemble striped bass, with whom they are related, in general appearance. Juvenile sea bass range in coloration from light brown to chocolate brown while older and larger specimens are usually much darker being jet black or a dark grey. The sea bass's tail and spine-tipped dorsal fin are usually tipped with white, and the

record for black sea bass stands at 9-pound, 8-ounces and was taken off Virginia. New Jersey's current state record for sea bass is 8-pounds, 2-ounces.

Bait and Lures

Part of the angling community's attraction to sea bass is the species' tendency to aggressively attack

large scales found along the body give the bass's flank a honeycomb appearance. Anglers should beware of the sharp spines located on the dorsal fin and centered on the gill plate which sea bass often use to inflict injury with when an angler is trying to unhook this fish.

Though capable of reaching 10-pounds, sea bass average around 1-pound. Anglers refer to large, male sea bass as "humpbacks" because of the apparent rise in the back of a large sea bass's head. The current all-tackle

and devour a wide variety of baits. Though sea bass will eagerly hit mussels, sea worms, shrimp, shedder crabs, green crabs, fiddler crabs, mackerel strips, and killies, clam and squid are the top two sea bass baits. Note that live killies are arguably the top bait for humpback sea bass; small sea bass tend to ignore killies.

One good aspect about clams and squid is they won't put a dent in your wallet like sea worms will. When fishing clams, most anglers employ surf clams rather than other varieties

Note how sea bass resemble a striped bass in general appearance. When unhooking these fish, anglers should watch the fish's spine-tipped dorsal fin and gill plates (center).

because surf clams are larger, providing more baits, and sea bass seem to prefer the surf clams' scent. Anglers typically cut 2 or 3-inch strips from the surf clam's tough tongue part being sure to include pieces of the clam's belly and

Baits aimed at humpback sea bass, such as live killies and large clam belly strips, are better used on the inshore wreck sites. Within the bay, small sea bass will aggressively attack these offerings but will eventually clean

ligaments. These provide a powerful scent and slight flutter that drive sea bass crazy.

Squid is the other popular sea bass bait. Though most anglers stick to squid strips, similar in size to the previously mentioned clam strips, anglers who use the squid's head or a long, fat tentacle from a large squid catch sea bass also. With sea bass fishing, one does not have to worry about cutting a strip too long because sea bass possess very large mouths, which allows them to inhale a strip that some anglers would think was too big. When fishing squid strips, many anglers dye their strips because sea bass seem attracted by color. Red, yellow or orange are the usual dye jobs for squid, and many anglers claim great results.

you out of bait due to their large and hungry numbers.

Unlike tog, sea bass will hit small lures. When over rough bottom, near rocks or over an inshore wreck, lightly jig either a $1/2$-ounce bucktail, 007 diamond jig or a 1/8-ounce Bridgeport jig along the bottom fluke fishing style. Sea bass will often attack the small lure mistaking it for a small baitfish. Lures are especially effective for humpback sea bass.

Tackle and Techniques

Around Barnegat Bay, the two most widely used rigs for sea bass are the standard fish-finder rig and hi-lo rig. Some anglers employ the regular

Note the pronounced "hump" behind this large sea bass's head that gives the fish its "humpback" nickname. Anglers can frequently find humpback sea bass around inshore wreck sites or occasionally near the inlet's rock piles (center).

fish-finder rig, equipped with either a #6 Virginia, a #4 Sprout hook, or a #6 English wide gap hook, to catch sea bass while drifting over rough bottom.

Most anglers prefer to employ the standard hi-lo rig because sea bass schools are usually thick and the likelihood of a double-header are good. Employ the same hi-lo rig you would for fluke or weakfish but substitute whatever hook you are using with the Virginia, Sprout or English wide gap hooks previously mentioned. Thanks to a sea bass's large mouth, one can get away with using a somewhat large hook.

Regardless of whatever rig you select, make sure to double hook your baits because sea bass will easily tear a single-hooked bait off a hook with minimal effort. Try threading the strip up the hook's shank to avoid a quick steal.

When fishing these baits, bounce your sinker on the bottom so the bait either flutters, jiggles or has some sort of movement because sea bass often find a slightly moving bait irresistible.

Where and When

Sea bass occur from Cape Cod, Massachusetts to Cape Canaveral, Florida, with the population heavily concentrated between New York and North Carolina. They are bottom dwellers that inhabit rock piles, clam beds, mussel beds, sod banks, bridges, pilings, bulkheads, wreck sites, shell bottoms, and other rough type bottoms. They tend to favor deep water.

Though they prefer their water on the cool side, sea bass are active in water 45 to 75 degrees. Sea bass usually come inshore during early May. They hold at wreck sites and rough bottomed locations in the inshore waters off Barnegat Light, providing anglers with great fishing until around mid-June. During the summer, anglers poking around wreck sites will catch sea bass but not as frequently as during the late spring. Around mid-August, small sea bass become very noticeable throughout the bay, holding near its rock piles, rough bottom and bridges. These small fish provide fluke anglers with plenty of action, and anglers will occasionally be surprised with a few nice size sea bass. Around late September, sea bass begin to leave the bay for their offshore wintering grounds. Along the way, they stop at the same inshore wreck sites and rough bottom they were at during the late spring. Anglers fishing these inshore locations, where the water temperature is cooler than the bay's water, can usually have a productive day with large fish.

Where in the Bay

Looking around the upper bay for potential sea bass holding locations, anglers should investigate the Point Pleasant Canal's mouth. The canal itself usually holds a large sea bass population near its bridge, sides and along its rough bottomed areas. Many of these fish will hold near the canal's southern mouth so anglers operating in the bay's Bay Head area can partake in some decent sea bass fishing. Anglers should focus on the channel that flows from the canal towards the Mantoloking Bridge. The edges of this channel and the deep water located near Dales Point typically produce good sea bass catches. Though these fish tend to be small, not being as large as the bass found on an inshore wreck, you will occasionally luck into an 8-incher while the rest will provide some plain old fun.

South from the Point Pleasant Canal area, the Mantoloking Bridge is a good place to check for sea bass. Look near the bridge's foundations and in the deeper parts of the bridge's channel. Sea bass tend to prefer holding in the channel's deeper parts, and anglers drifting a hi-lo rig baited with clam or squid strips will get a nice bend put into their rod by a resident sea bass. Similar to the canal area, the sea bass you are likely to encounter at the Mantoloking Bridge are 5-inch rascals, though there are some respectable 8-inchers swimming about here.

South from the Mantoloking Bridge, anglers drifting the channel that winds past Kettle Creek and Silver Bay typically catch sea bass as the random catch of a fluke expedition. The better fishing here is generally in the mid-August to September time frame.

Further south, anglers will find sea bass holding near the Route 37 Bridge, especially where it spans to Pelican Island. Here, anglers fishing close to the bridge's foundations and drifting baits in the deep water near the bridge often meet the bridge's resident sea bass population. If the bridge is devoid of any action, try the channel that winds beneath the Route 37 Bridge south towards the Toms River. Anglers will usually find sea bass holding near the channel's depressions.

An unsuspecting sea bass spot that lies south of the Toms River is Tices Shoals. Later during the summer, anglers fishing for weakfish at Tices Shoals will catch sea bass. To most anglers fishing the shoals, the small sea bass are nuisances that raid the shedder crab and squid baits employed for weakfish. At Tices Shoals, sea bass tend to occur in the deeper water near where the bottom rises up to the flat.

As you near Barnegat Inlet, you begin to find greater sea bass concentrations. For anglers investigating Oyster Creek Channel, the channel's rough bottom areas located particularly at its eastern mouth, and coral beds typically hold hold hordes of small sea bass during the late summer. The channel's western mouth, near buoy #67, is another usually productive sea bass area for anglers employing chum for blowfish or weakfish.

Barnegat Inlet is where Barnegat Bay's larger sea bass tend to abound. With its deep holes, cold ocean water, rock piles and strong currents, the inlet is as close to the oceanic environment, that sea bass prefer, as one will find in the Barnegat Bay area. Not surprisingly, anglers can catch sea bass at the inlet throughout the summer — even at its hottest. The inlet's main channel, Meyers Hole, and the sod banks are very good locations to find sea bass. The rocks that line the lighthouse and merge into the south jetty yield sea bass to anglers employing clams or squid. At the north jetty, anglers will catch sea bass if they fish tight to the jetty's rocks. Curiously, the inlet's main channel, where it flows parallel to the north jetty, rarely yields sea bass to drifters.

Looking back at the bay for sea bass, the rough bottom located at Double Creek Channel's eastern end plays host to a significant sea bass population during the late summer. Look near the sod banks that border the channel's eastern flank and where the channel heads past Clam Island.

Large, humpback sea bass are mostly found at the inshore wrecks that lie off Barnegat Inlet. Even if you cannot anchor up to an actual piece of the wreck, drifting the wreck site's perimeter is always a plausible avenue to catching humpback sea bass while also perhaps putting a dent in the fluke population that always takes up residence around the wreck site's immediate area. Try drifting a live killie or a whole, freshly shucked surf clam to catch humpback sea bass at wreck sites.

Anglers can also find sea bass at inshore rough bottom areas, particularly Harvey Cedars Lump, and at shellfish beds, a few located off Island Beach, in less than 50-feet of water.

Not to preach, but anglers should practice some conservation especially when fishing for sea bass. They are aggressive fish that will keep coming like bluefish. So rather than keep 50 or 80 fish, trying keeping only several and throw the rest back for a second trip.

Chapter 6
Winter Flounder

Once the water temperature plummets during the late autumn, Barnegat Bay becomes a ghost town as its species either migrate or hibernate. As everything is preparing for Old Man Winter, winter flounder begin to take up residence in the bay as the water cools to their liking.

Winter flounder fishing brackets the year for many anglers who are itchin' to fish and like the flounder's delicious flesh. Prior to the law that dictates a closed season during the winter, diehards enjoyed fishing throughout the year by targeting winter flounder during the dead months of January and February. Now, the flounder fishing season opens every March 1st, but most anglers interested in winter flounder typically take a stab at this fish during April. After a summer filled with crabs, weakfish and fluke, anglers once again settle on catching winter flounder during late November and early December.

Also known as "black-back," "muddab," "flatfish" or more commonly "flounder," winter flounder (Pseudopleuronectes americanus) resemble fluke in general appearance.

The most notable difference between the two fish is found in the head area. A winter flounder possesses a pointy-looking head tipped with a small, rubbery-lipped, O-shaped mouth that it uses to nip clams poking out from mud and to suck up sea worms. Fluke, on the other hand, possess a much broader head equipped with a tooth-filled mouth that is significantly larger in proportion to its body than a flounder's mouth is to its body. Also, a winter flounder's eyes are situated on the right side of its body while a fluke's eyes lie on the fluke's left side.

Though they do not reach the tackle-busting size of fluke, winter flounder can make a good account of themselves when faced with light tackle. Within Barnegat Bay, winter flounder average around 1-pound in weight and 10-inches long. Larger ones will tip the scales at a little over 2-pounds, being about 16-inches long. Flounder of all-tackle record proportions probably visit Barnegat Bay each year. Back in 1981, a lucky angler fishing near Gulf Point caught a 3-pound, 4-ounce winter flounder that was soon recognized as the state record for the species.

Winter flounder bring smiles to many anglers' faces because for many, winter flounder mark the beginning of another fishing season (opposite).

Flounder Fare

The top winter flounder baits are: sea worms, clams, and mussels. Nightcrawlers, squid and corn also make the bait list but are more frequently used by anglers either in desperation or when they have many hungry flounder in their chum slick, and the flounder are whipped up so they're not exactly picky.

To a seasoned flounder angler, "sea worms" can pertain to several types of marine worms that flounder prey upon within the marine environment. Bloodworms and sandworms are the two more popularly employed worms while tapeworms and greenworms find use among more savvy anglers. Similar to using grass shrimp for weakfish, smart anglers use tapeworms or greenworms because these are the worms the bay's flounder population will sometimes feed exclusively upon.

Some anglers prefer to use small 1-inch worm pieces so a flounder can easily fit the bait into its small mouth and so the angler can get many baits from the expensive worm. These anglers are on the right track after they have attracted aggressive flounder with a chum slick. Once flounder are whipped up into feeding, why waste bait?

Other flounder anglers employ longer 3-inch pieces, hooking the worm so it drapes off the hook. Longer worms appear more natural than a small 1-inch chunk. In the marine environment, flounder have probably seen many 3 or 5-inch worms but not too many 1-inch specimens that are as wide as they are long. Do not think that flounder will have a tough time eating long worm baits. Small "coaster" flounder have little problem sucking up a 3-inch worm. Imagine what a 16-inch flounder can suck up. Some anglers go a little ape and start hooking whole 7-inch worms on their hooks. Don't laugh, I have seen these people catch some nice size flounder.

When using a worm for bait, cut the worm as you use it rather than immediately cutting it into three or four pieces. Cutting the worm up in one shot causes it to leak most of its fluids resulting in the angler fishing with a piece devoid of its scent.

If you want to save time and cut worms up at once, cut them over a small pail and leave the severed worm or worm pieces in the pail. Rather than cut a worm on a cutting board and have its bodily fluids leak out and soak into the wood, the pail will capture all the liquids. The worm or worms left in the pail will stay moist and juicy, marinating in the liquids while washed out baits can get basted and re-used if need be.

In any event, whatever size worm you choose to employ, keep your worms in good condition. Keep them away from the sun's drying effects, in a cool, dark location such as an ice-filled chest. If kept cool with the tackle store-supplied seaweed, the worms will stay lively, even trying to pinch you with their mandibles. The lively and juicier the better, because the worm's movement and scent is what will catch a flounder's eye.

When it comes to using sea worms for bait, flounder will sometimes bust your chops. The times when you pinch pennies at the tackle store and go fishing with a dozen worms is when the flounder eat like there is no tomorrow.

When you paint the town red and go fishing with 3 dozen worms, this is when the trip flops and you end up using only 4 worms. Rather than toss the worms, you can keep them for a week or so by storing them in a cool, dark location. Refrigerators are ideal for worms, but you can get away with placing them in the garage or basement.

catch several flounder from one location and all of their stomachs are filled with severed clam feet, then you can bet that within the area you were fishing, there was a clam bed. Clams, and also mussels, are a good bait to use when the water is cloudy because flounder will usually key onto this lightly colored bait that stands out in the dim water.

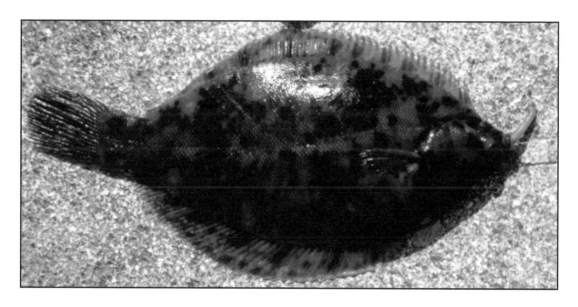

Keep them a tad moist, but not too moist, by placing seaweed atop them. Line the worms' container with newspaper to soak up excess moisture. Without the newspaper, too much moisture will kill sea worms.

Clams are closely behind bloodworms and sandworms as the most popularly fished flounder bait. They are a very potent bait, especially when you employ them near clam beds and other areas where flounder prey upon clams. An easy way to figure out where clam beds are, besides looking at a map, is to look inside a flounder's stomach when filleting it. Though flounder do move with the tides, if you

There are to my knowledge, three varieties of clams that area flounder anglers use: steamer, hardshell and surf. Steamer clams are the white, brittle-shelled clams that constantly expose their fat, tubular body part and are commonly sold in supermarkets. Many anglers fish the tubular body part because that's the part flounder nip off when they see it sticking out from the bottom.

Hardshell clams are probably the most prevalent clam in the bay and logically make a good flounder bait because flounder frequently feed on these mollusks. Fashioning small $1/2$-inch to 2-inch long strips, that are about

Similar to fluke, flounder possess the ability to darken or lighten their coloration to camouflage themselves against the bottom. Depending on the bottom, flounder will range in color from a dark chocolate brown to a light reddish-clay. Lighter colored flounder will usually sport dark splotches to better blend them with an unevenly toned bottom. Their undersides are white (center).

¹/₄-inch wide, from a freshly shucked hardshell clam are good baits to fish near a clam bed or down tide from a clam raker.

Surf clams also grace many a hook. Though some anglers prefer to fish hardshell clams, many agree that freshly shucked surf clams cut into small strips make excellent flounder fare. Note that live clams are heads above frozen ones because a strip cut from a surf clam still pulsating has a

blowfish — twice rather than once so a quick suck from a flounder won't fleece your hook.

Remember to change clam baits that have seen extended duty for a fresh, juicy strip. You can, however, take the washed-out strips, whose juices have been leached out by the water, and marinate them with some bloodworm blood, other worm juices or clam juice via the previously mentioned "juice pail" scheme. Some anglers even

Surf clams are a staple item employed by winter flounder anglers. Whole clams are cracked and used as chum while others are shucked and fashioned into strip baits. Many anglers shuck surf clams and grind them up for their own homemade chum recipes (center).

texture and scent, besides movement, that frozen clams cannot match.

When fishing clams in general, thin strips are the key. The length of the strips, similar to the length of worms, can vary from ¹/₂-inch all the way to 3 or 4-inches but the strip's width should only be about ¹/₄-inch. I do not know if thin strips flutter in the water or just look more appetizing, but I have noticed better action when fishing thin strips. Hook a strip similar to the way you would hook a squid strip for sea bass or

marinate clams the night before fishing in either yellow or red food dye because flounder are very keen on colors, especially yellow and red. Many anglers claim yellow food coloring makes a clam strip appear as a bright, healthy clam foot sticking from the bottom while red food coloring makes a clam strip appear as a sea worm. Dyeing baits can be very effective at times.

Besides clams, the other mollusk that sees considerable action are mussels. Many anglers make the mistake of seeing mussels in a single light. Mussels, both blue and ribbed, can figure into the flounder fishing equation in two ways. One, they make excellent chum whether finely grounded up or simply smashed and dropped overboard. Two, larger mussels are tantalizing when threaded on a hook and fished whole. Because mussels are a primary food for many Barnegat Bay flounder, anglers who use mussels for bait usually catch flounder even on lean days.

Blue mussels are the black, smooth-shelled mollusks that attach themselves to bridge foundations, pilings, bulkheadings, and rocks along the beach. Ribbed mussels get their name from their shells, which possess lengthwise ribs. These type mussels are often a mud brown color and can be found along the bay's muddy, salt marsh banks.

When using mussels for bait, remove the mussel from its shell via a nutcracker or knife and thread the entire mussel on the hook, employing the hook's shank if necessary. Make sure the mussel is firmly hooked, because they tend to be much softer than clams, and part hangs off the hook so it will swing in the water.

Many areas within the bay that play host to flounder are also home to mussels. It is no surprise that locations such as the Mantoloking Bridge and the Pelican Island Bridge whose foundations sport blue mussels also attract flounder.

Nightcrawlers, squid and corn are commonly brought out when anglers are confronted by many flounder and don't have enough premium bait. In such scenarios, anglers extend their bait supply by using these baits.

Many anglers employ the relatively inexpensive and easy to dig nightcrawler or garden worm in place of the more expensive sea worm. Note that saltwater will kill nightcrawlers and garden worms rather quickly, so you should bring plenty of worms on a fishing trip if you plan to rely on them for bait.

Tiny, thin squid strips, about an inch long and a $1/4$-inch wide, have been known to work when used in a chum slick accentuated with rice. Similar to clam strips, some anglers dye their squid yellow or red to give it a clam or worm look.

Regular canned corn, sometimes used as chum, will also account for flounder who mistake the kernels for clams poking out from the mud. Before breaking out the can opener, you should have a few flounder in your chum slick. As mentioned earlier, these three baits should be used as your reserves and not as you main baits. As a side note, some old salts will use the regular supermarket bay scallops or oysters as bait often with good results.

Tacklin' Flounder

Winter flounder fishing is a light tackle affair. Light action graphite rods, small freshwater style conventional or spinning reels, and light line are the ideal components to a fun and successful day of winter flounder fishing.

The better rods to use for flounder are $5^1/2$ to 7-foot long, light action to medium action stiff-tipped graphite rods. A stiff-tipped graphite rod is particularly good for flounder fishing because the rod's inherent sensitivity allows you to feel the slight bumps that flounder often register when sucking up a baited hook. This type rod also makes it easier to set the hook quickly before the flounder can spit the hook out.

Winter flounder anglers employ both spinning reels and conventional reels. To me, it's a matter of personal preference over which to fish with, though shorebound anglers needing to cast a long distance should consider using a spinning reel unless being adept at using a conventional reel. Whichever type reel you decide upon, the light freshwater reels, that are

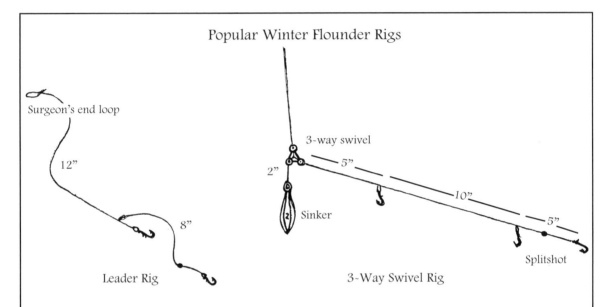

Popular Winter Flounder Rigs

Surgeon's end loop

12"

8"

Leader Rig

3-way swivel

2"

5"

Sinker

10"

5"

Splitshot

3-Way Swivel Rig

After selecting a hook, you can construct a flounder leader rig which is connected to a fish-finder or drooper loop rig. Snell the hook to a 12-inch piece of 10 to 15-pound leader material.

Many anglers, particularly ones fishing from boats, where they can net a "doubleheader," employ light line and thus light leader, sometimes as light as 8-pounds, so they can use less weight to hold bottom and can really feel a flounder's slight nudges. Some anglers fishing from high bulkheadings or near bridges where they may have to lift a "doubleheader" of two fat flounder up from the water while potentially rubbing up against a sharp, barnacle-encrusted piling, employ leader as heavy as 20-pounds.

After snelling the hook, take a single colored bead (red, yellow or pearl) and run the leader through the bead's hole, letting the bead rest against the hook. Tie a surgeon's end loop at the leader's other end. About 2-inches above the hook, tie a drooper loop. Take a second hook and snell it to an 8-inch piece of leader and tie the leader's other end to the drooper loop. You should now have a double hook rig. On a two-hook rig, I often experiment with placing a single split shot about an inch above one hook to keep it closer to the bottom than the other. Sometimes flounder will hit baits very close to or on the bottom while other times they don't seem to care. Experimentation is often the key with flounder because they are not the easy targets some anglers make them out to be. Anglers then attach the leader "rig" to either a fish-finder, 3-way swivel or drooper loop rig. (left)

A popular "stand alone" rig consists of a 3-way swivel, a 20-inch piece of leader material, 3 hooks, a splitshot, and a sinker. Take one hook, place it 5-inches from the leader's end and tie a drooper loop so the hook lays right in the loop's middle. Take the second hook and attach it to the leader 10-inches down from the first hook via a drooper loop. Snell the third hook at the leader's end and attach the splitshot about 2-inches before the third hook. You then connect the leader's other end to the 3-way swivel's bottom right eyelet, a sinker to 2-inches of line to the other bottom eyelet and the line from your reel to the top eyelet (right).

good for flounder fishing in the bay. They're lightweight, typically have good drag systems and pack more than enough line to tackle the biggest flounder you are ever likely to encounter in the bay.

For flounder fishing within the bay, 8 or 10-pound test is sufficient. Anglers planning to fish from a bulkhead or bridge should consider employing a reel with greater line capacity to compensate for the increased distance between their position and the water. You might also want to try heavier line so you can lift a doubleheader of fat flounder out of the water without a net. Anglers operating from a boat who have the luxury of a net to handle a doubleheader can go as light as 6-pound test. The lighter line allows for greater sensitivity and less weight.

Rigs

Rigs aimed at winter flounder commonly consist of two to four hooks so anglers have a shot at catching multiple flounder in a single shot. A single yellow, red or pearl bead is usually set above each hook to attract flounder to the bait. Anglers use yellow beads when fishing near clam beds because the bead appears as a clam poking through the bottom and quickly grabs a flounder's eye. A red bead mimics a worm head while a pearl bead appears as a rice grain, which many anglers employ as chum.

You may want to tie one rig with colored leader material. This leader material is typically yellow or red, and many anglers report good results when using colored rigs. Frankly,

it's smart to carry a colored rig and switch to it if your fishing expedition is going down hill fast. Similar to weakfish, flounder will on some days respond to any bait or rig ever associated with their kind. On other days, they will totally avoid the standard fare, only hitting certain baits and rigs. When not getting any bites, the anglers who break out a yellow or red rig are the ones who consistently catch flounder.

Winter flounder anglers, unlike tog enthusiasts, do not have a hook they all champion. While the Virginia style hook is widely regarded as the tog hook, the Chestertown, sproat, aberdeen, bait holder, English wide gap and beak style hooks all have popular followings with the former two hook types being the more popular choices among flounder anglers. Many anglers opt for the Chestertowns, which possess a very thin gap for the flounder's small mouth and a long shank for easy hook removal. Popular sizes range from a #6 to #10. Other anglers prefer the #6 to #8 Mustad #3399 Sprout model, the #10 thin wire Aberdeen model, or typically the #8 beak style hook. I like to fish with either a #9 or #10 Eagle Claw baitholder hook. These size hooks are small enough to conceal inside a bait and hook a flounder that wants a taste but not a meal. Whatever hook you settle upon, keep in mind that flounder have small mouths. While they can be aggressive at times, and will literally inhale a hook, there are significantly more times when small hooks save the day because flounder are lethargic and nip at baits.

The more popular rigs associated with winter flounder fishing

Did You Know?

The current all-tackle record for winter flounder stands at 7-pounds and was set off Fire Island, New York in 1986.

A 5-pound, 11-ounce specimen caught off Barnegat Light is the current New Jersey State record for the species.

are the fish-finder, 3-way swivel and drooper loop. For the fish-finder, simply connect the previously described leader rig to a swivel that is tied before a regular fish finder. Many anglers prefer using a fish-finder rig because it allows them to feel a flounder's delicate bite. If you are the type that likes to jig your bait rather than dead-stick it, a fish-finder rig will give you greater sensitivity. Use the fish-finder rig if you're an angler who likes to drift for flounder on extremely placid days.

The 3-way swivel rig is probably the most the commonly employed rig amongst anglers. Many anglers prefer using 3-way swivel rigs because the swivel helps keep the rig and sinker separate. When fishing with this rig, anglers connect the previously described leader rig to the swivel's bottom right eyelet, a clip with an attached sinker to the bottom left eyelet while the line from the reel is attached to the swivel's top eyelet.

The drooper loop rig's great attraction amongst anglers is that it contains no hardware. For the drooper loop rig, tie a small surgeon's end loop at the end of your main line. Tie a drooper loop $1/2$ to 2-inches above this end loop. Attach a sinker to the end loop while attaching the previously described leader rig to the drooper loop.

Sinkers for bay flounder fishing, typically bank or coin style, usually range from an ounce to 4-ounces. Anglers commonly employ 1 or 2-ounces when fishing with light line in a relatively sheltered area like the Toms River or Double Creek Channel. Four ouncers see action when one fishes with heavy line, which causes water drag, or when fishing in a deep location that

enjoys a swift current, such as Meyers Hole. Many anglers employ heavier than needed weight to stir up the bottom with rhythmic sinker bouncing. Painting sinkers yellow, red, orange, or white is popular among flounder fishing clans because these colors supposedly mimic the small organism flounder feed upon. Flounder are known to be attracted by colors, and anglers who employ painted sinkers usually claim good results.

Chum Some

Chumming is synonymous with winter flounder fishing because it gets sluggish flounder going when the water is cold. With chum, it's possible to attract winter flounder to one's position from a considerable distance. Chumming for flounder usually takes three forms which are sometimes used alone or in combination with each other: cracked mollusks, chum pots and simply stirring up the bottom with a long, stout stick.

Rather than fiddlin' with chum pots and messy chum logs, some anglers simply chum with whole hardshell clams, steamers, or mussels. First, they bludgeon the mollusks with a hard object and then drop the cracked shelled mollusks right where they plan to drop their baited hooks. They initially toss about five to six clams or a dozen mussels when starting their slick. Flounder will smell the mollusks' bodily fluids and follow the scent to the cracked shells. Many anglers employ mollusks simply because the freshness factor and scent from crushed clams or mussels will not get washed-out by a swift current as quickly as ground chum. When flounder start hitting the angler's baited hooks, he or she then cuts down to two to three clams or six mussels. When fishing in cold water, during March and early April, you may want to chum heavy to roust the flounder from their dormancy.

Chum to most winter flounder anglers translates into the frozen,

together or use them to add bulk to a homemade chum concoction. Most anglers create their chum by taking blue or ribbed mussels and smashing them in a bucket with a 2x4. The crushed mussels are then placed in a chum pot as is, or mixed with pet food or boiled rice. Pet food is a good chum

ground variety sold by tackle stores in 10-inch logs. Many anglers favor chum logs because they are readily available and offer little hassle. To use a chum log, simply unwrap the log, pop it in either a weighted chum pot, sold at tackle stores, or a plastic, gallon milk jug with holes poked in it and bricks attached for weight and you are ready to chum. If you plan on relying solely on chum logs for your slick, keep in mind that a swift current tends to wash logs out pretty quickly so be prepared to bring enough.

Many anglers, including myself, either avoid chum logs all

additive because its fineness causes it to become quickly washed out of the chum pot, which quickly starts the slick, while the crushed mussels keep the chum slick going. The generic tuna or seafood flavored cat food, in particular, is cheap and makes a great addition to crushed mussels. Boiled rice mixed in with crushed mussels and pet food works well. Cooked rice, perhaps from its plumpness and brighter coloration, seems to work better than uncooked rice. Whenever you employ rice as chum, thread a single pearl bead above your hook. Flounder will often strike

"Doubleheaders" are common occurences when flounder fishing is in high gear. Make sure to bring plenty of bait because when flounder are on the bite, it is easy to go through a few dozen sea worms or clams (center).

hooks adorned with a pearl bead because it closely resembles the rice.

You can either place the chum in a weighted chum pot, designed specifically for chum logs, or you can place the brew in a plastic, gallon milk jug. Tie a few bricks, concrete-filled soda cans or a few 16-ounce "cod fishing" sinkers to the jug so it hugs the bottom. Right before you place the chum jug in the water, poke holes in it with an ice pick.

When fishing from a boat, a smart way to chum is with double anchors. Employing a single anchor often results in the boat swinging from left to right, which lessens the chum's effect because the boat's sweeps disperse the chum over a wider area. Single anchoring also results in having the boat vertically aligned where three anglers try to squeeze at the back of the boat. Double anchoring binds the boat to a single place and keeps it horizontally aligned with the area you are planning to chum. With horizontal alignment, three anglers can leisurely spread out.

Double anchoring for flounder is done the same way you would if anchoring for weakfish or bluefish chumming. Drive up current from a flounder holding spot, such as a channel ledge, drop the bow anchor first, let out a little line and then drop the second anchor from the stern. Let out anchor rope until the boat drifts towards the flounder spot. Tighten both ropes when the anchors have embedded themselves in the bottom, and the boat lays horizontally a tad up current from the flounder spot. Securely tie the ropes to the boat's cleats and start the chum slick.

If cracking mollusks for chum, start dropping them over the side. If using a chum pot, place it in the water off the up current side of the boat so the chum seeps underneath the boat towards the flounder spot. If you are fishing multiple chum pots, place them so they converge to form a large, unified stream. Give the chum pot(s) a shake near the surface so you can see the chum's width and direction. When chumming with a chum pot, take a hook or two and connect them to your chum pot. Flounder that have infiltrated the chum slick and gotten past your lines are caught many times by baited hooks attached to the chum pot.

Cooked or uncooked corn and rice is then tossed far up current and allowed to drift into the slick. Anglers employ yellow corn to mimic the yellow color and shape of the clam tongue tips that flounder love to dine upon while clear uncooked rice resembles small shrimp or arthropods.

Periodically during the day, while chum is seeping from the chum pot, anglers should stir up the bottom with a long, stout stick. Chum, whether homemade or store-bought cannot fully duplicate all the crustaceans, arthropods, immature mollusks, and other organisms that become released from the bottom and float down current every time an angler stirs the bottom with a long, stout stick, a mushroom style anchor or a long stick tipped with a plunger. As a side note, savvy anglers will typically anchor down current from a clam raker who, during his or her toils, will stir up the bottom doing the work for the angler.

Once you attract flounder to your position, back off on the chum a little to conserve it and to avoid feeding the flounder. If the bites slack off, which they often do in cold weather, go back to employing more chum.

Where and When

Winter flounder occur from the Gulf of the St. Lawrence River in Canada to the Chesapeake Bay. A bottom flounder. At some rivers and creeks, flounder will venture so far upriver that they eventually reach freshwater. Winter flounder will generally return to the same location each year making it easy for us anglers because once one locates a flounder honey hole, the hole will usually produce flounder year after year.

dweller, winter flounder prefer bottoms that are a muddy sand mixture. Not surprisingly, a wide variety of flounder prey, particularly sea worms, occur in this type of bottom. During the course of their movements, flounder will occur over bottoms composed of sand, soft mud or clay, concentrating themselves near a steady food supply. Ergo, clam beds, marsh banks, bridge pilings with attached mussels, eel grass forests, creek mouths, river mouths, and muddy sand bottomed locations typically hold winter

A cold water fish, winter flounder invade Barnegat Bay when the water temperature drops into the 40s during the late fall. Though flounder season opens legally on September 15th, the first significant catches are put together around late October, usually at the inlet. By mid to late November, anglers fishing throughout the bay begin putting together catches and flounder fishing usually picks up around late November and early December, depending upon the water temperature. Though the flounder

A nice winter flounder caught on a typical brisk flounder fishing day. Note the angler's warm clothes and ear muffs. Though targeting warm, sunny days for fishing, it's usually still very cold. Dress warm when seeking flounder because you want a double-header of fat winter flounder to push you to the limit and not Old Man Winter (center).

season closes December 31st, and stays closed until March 1st, diehard anglers catch flounder even during the bitter months of January and February by fishing the warm water discharge of the Oyster Creek Nuclear Power Plant. When the flounder season opens on March 1st, anglers find flounder generally receptive if they chum heavy on sunny, windless days at proven flounder holes. Mid-April to mid-May, when the water temperature approaches the high 40s to low 50s, is a very good time to catch flounder as they begin to gather themselves and move from the bay to their offshore summering grounds. Once May 31st rolls around, the winter flounder season closes until September 15th. Flounder do, however, maintain a year 'round presence within the inshore area. Anglers can usually find them near inshore wreck sites and within extremely deep parts of the bay, such as Meyers Hole or the 40-foot plus deep hole located directly in front of Barnegat Lighthouse.

Flounder are very sensitive to their environment. Tides, sun and rainfall all play a hand in whether flounder will bite or get lockjaw. During the fall, you will generally find the better flounder catches are produced during an incoming tide because the incoming water brings flounder into the bay and also brings the cooler, ocean water which stimulates flounder more than the relatively warm bay water. During the spring, better flounder catches are typically put together during an outgoing tide because flounder will become re-invigorated by the warm outgoing water, and they will use this tide to exit the bay.

In addition to tides, anglers should watch the weather when planning a fishing trip. Sunny, windless days are ideal days to fish for flounder. The warmth generated by a sunny day versus a cloudy day is often significant enough to trigger the flounder into feeding. Fishing a few days after the first spring warm front can be very productive because three or four days of warm weather can warm the water slightly causing flounder to shake their dormancy and seek food. Conversely, a windy, cloudy day often puts the clamps on any successful fishing. The water temperature either drops or stabilizes and anchoring a boat becomes a colossal feat. Besides not making the air temperature drop, windless days make it so you don't have to break your neck trying to set anchors and having the boat swinging all over.

Besides avoiding cold, windy days, you also don't want to necessarily fish right after a torrential rain storm. The waters at many locations within the bay, such as the Metedeconk River and the Toms River, occasionally become murky after a strong rain. Whether it's because the water salinity changes or the tiny dirt particles aggravate their gills, flounder do not typically respond well to murky waters. Note this does not pertain to stirring up the bottom for chumming purposes because that merely dislodges flounder prey from the bottom and is confined to a relatively small area. When confronted by murky water, try fishing a location that constantly gets flushed out such as Barnegat Inlet or wait for the water to clear before spending money on bait and chum.

Upper Bay

During the early winter and spring, the Point Pleasant Canal's mouth is a good place to fish for winter flounder because it is one of only two winter flounder access points to Barnegat Bay. Here, anglers pick winter flounder off as they move into the bay during the fall and as they exit the bay during the spring. A few depressions and holes located directly south from the canal are productive. Fishing too close to the mouth is a little hectic during a moving tide because of the strong current.

South of the canal, many anglers do well when fishing the waters around Herring Island. The channel that winds past the island's eastern side is a favorite flounder area because flounder get the benefit of a deep channel to lay in on cold, cloudy days, and shallows, near the island, to prowl when it's sunny and warm. Many anglers fish the edges of this channel because flounder will leave the channel's deep pockets during a high tide to venture up into the shallows which flood during the high tide. The shallows' relatively warm water and prey attract flounder. Thus, anglers fishing the channel's slope can pick off flounder as they migrate into the shallows during a high tide and when the flounder head back to the deep water during a receding tide. Anglers should note that a good portion of flounder do not actually venture into the shallows. Usually, they stay along the channel's slope where they can pick off debris as it is brought in by the current.

A tad south from Herring Island lies a very popular winter flounder location: the Mantoloking Bridge. Though the flounder here only average around 10-inches, being not as large as the average flounder taken from the inlet, the Mantoloking Bridge houses a healthy flounder population. At the Mantoloking Bridge, flounder typically hang near the bridge because the mussels and other organisms that cling to the bridge's foundation provide flounder with a steady food supply. Plus, the bridge is close to the Point Pleasant Canal where flounder can have quick access to the Manasquan River and from there to the ocean.

When fishing the Mantoloking Bridge, many anglers target the channel that winds beneath the bridge, paying close attention to the channel's edges and the depressions that occur within the channel itself. Perhaps the better flounder spots at the bridge are the several deep holes located a hundred yards or so south from the bridge's southern side. These holes range in depth from 8-feet to about 15-feet and are relatively small, being able to accommodate about two boats depending upon how well people anchor. The trick to anchor at these holes is to have your chum seep into the hole and not in the surrounding shallows. Anglers typically fish either a hole's edge or dead over it. Usually, being dead over it so you fish the hole's deepest reaches accounts for good flounder catches. If you cannot locate a hole or come upon an "early bird" boat straddling a hole, fear not because the channel that branches off from the main bridge channel holds flounder.

Anglers looking for winter flounder south from the Mantoloking Bridge should consider Chadwick Beach. Chadwick Beach, located on the

Did You Know: Commercial fishermen have taken winter flounder as large as 8-pounds from New England waters. Locals call these size flounder "snowshoes."

bay side of Normandy Beach Township, has a deep hole off of it that usually holds flounder. This deep water plummets to over 15-feet in some parts and is usually productive during the spring. Late March to mid-April are typically good times to fish, while anglers poking around here in early December will usually catch fish. If you take time to investigate this location you should make out fairly well with its flounder population.

Southward from Chadwick Beach lies the Toms River which has many locations near it and within itself that host flounder. Flounder anglers operating around the Toms River area generally concentrate their efforts at Pelican Island and within the Toms River.

When considering Pelican Island for flounder, one particularly good area to investigate is a deep pocket located northwest of the island. This pocket, situated north of the small channel that runs parallel to the Route 37 Bridge is a relatively large area that can accommodate a few boats without anyone feeling crowded. Not too far from this pocket lies the bridge that connects Pelican Island to Island Beach which is a favorite area for flounder anglers. Usually, the best flounder fishing takes place at the bridge's eastern end before it hits Island Beach. Anglers typically fish the nearby channel's deep waters at low tide or on cloudy days and then fish the shallower water close to the bridge at high tide and on sunny days. Anglers operating a boat can take advantage of the flounder opportunities available at Pelican Island's southwestern shores. Here, the water drops below the 10-foot mark a

short distance from the island. Look for flounder to hold near the uneven bottom and deep pockets. Here, you will typically find isolated flounder schools where you will catch several and then have to relocate to catch several more rather than anchoring in one spot and hitting the "mother load."

Anglers fishing the Pelican Island area don't have to travel too far if they want to try another location to increase their flounder score because southwest from Pelican Island lies the Toms River. For the most part, the river's mouth and the Island Heights area, located in the river's center, generally offer the best fishing. Popular areas to fish at the mouth are Coates Point and Good Luck Point. Though the river's mouth tends to have a seemingly even bottom, it does possess a few contours and dips that require some investigation to locate, but once you locate them you should have a steady "honey hole" for the year.

Westward into the river, Dillon Creek's mouth is good for flounder and so is the water near Long Point, especially if anglers fish where the water deepens just off the Point. Anglers will also catch healthy numbers of flounder in the Island Heights area, particularly in the vicinity of the Yacht Club Pier. Also, the waters off the eastern tip of Mill Creek's mouth is usually good to fish during the early spring. The Island Heights area though, generally offers better fishing when considering the river's interior for flounder. As one heads westward into the river, flounder tend to become smaller and sparser.

Lower Bay

To many anglers, Barnegat Bay's lower expanse is the place for consistent flounder catches and a crack at the largest flounder that swim the bay.

Oyster Creek, flowing in the shadow of the nuclear power plant, is an early season hotspot. As soon as March 1st rolls around, anglers usually start catching flounder, attracted by the power plant's warm water discharge, from the Route 9 Bridge.

Oyster Creek Channel is one of the bay's more well-known winter flounder spots. Part of Oyster Creek's attraction to flounder, besides being a route directly to the inlet where the bulk of the bay's flounder enter the bay, is it cuts through an eel grass flat. Flounder prefer hanging near eel grass flats because the flats house a variety of prey and are usually rooted in the soft sandy mud bottom that flounder prefer to inhabit.

Oyster Creek's western mouth, around buoy #67, produces good winter flounder catches for anglers who anchor up current from the small pockets and contours that dot this area. Following Oyster Creek eastward towards the inlet, the entire channel provides opportunities for flounder anglers. One fishing along the channel's edge, so chum seeps into a deep pocket or contour located within the channel, will catch flounder. One particularly good area to try for flounder in the channel is where the channel swings north into the eel grass flat and then slides back south before it makes a sharp southernly turn towards the inlet.

Shortly after this turn, Oyster Creek flows into Barnegat Inlet.

Barnegat Inlet is a very popular area to catch flounder during late October and November, when flounder first enter the bay, and again during late April and early May, when most of the bay's last flounder leave for their summering grounds. Within the inlet, there are two main spots that flounder anglers focus upon: the inlet's main channel and Meyers Hole.

Anglers can simply head north into the inlet's main channel and fish along its edges, within sight of Old Barney (the Barnegat Lighthouse) and catch flounder. Part of the main channel is sandy so you want to scout around and find the nice muddy sand mixture or plain mud and avoid fishing over the underwater Sahara. The channel's northern side, where it borders a sand bar, is one example of an area where one doesn't want to necessarily concentrate their efforts. Most anglers fish where the channel splits, going north towards the ocean and south towards Meyers Hole and Long Beach Island. At this junction point, and a tad south from it, anglers will find good flounder fishing during the spring. Remember that it's illegal to anchor within the channel, so your boat obstructs boat traffic; pay extra attention to your anchoring job because a Coast Guard base is located a stone's throw away from this junction.

Nestled within the inlet south of the main channel lies Meyers Hole the mecca for the many winter flounder anglers who ply the lower bay. The fact that the occasional angler will catch a winter flounder during the summer months while fishing Meyers Hole attests to the species' attraction for this spot.

Meyers Hole is a large, deep mud and muddy sand coated depression that's about 32-feet at its deepest. The deeper parts of the hole located near a small marsh island, where the previously mentioned channel winds past, is where many anglers take their fish. Even though the fishing here is often superb, the location of this spot in relation to the channel may present problems on heavily traveled days. The channel that winds through the deep areas near the island marks one of the only routes many boats moored at Barnegat Light use to reach the inlet. On busy days, it is often more enjoyable to stay away from the island area and fish the western and northwestern parts of the hole than contend with the boat hoopla.

One of the beauties of Meyers Hole is it is large enough to fit several boats without anyone feeling congested. On the other hand, it is small enough so the flounder are concentrated to keep you from having to search an immense area for fish. If the fishing is slow, gradually move the boat to where the hole slopes up or down until you find the depth the flounder are holding at. Usually, experimentation with your location is the only difference between a good day fishing and a bad one. One other benefit about Meyers Hole is that it is shielded from the wind that often kicks up during a typical fall or spring day. A howling wind that can complicate anchoring in unshielded areas is usually nullified by the land and shallows surrounding the hole. Additionally, Meyers Hole is close to Oyster Creek and Double Creek Channels so anglers can try Meyers Hole and in the unlikely event the flounder

have lockjaw, anglers can fish other locations without having to trek across the bay.

To the south of Meyers Hole, behind Long Beach Island, flows Double Creek Channel which is arguably one of the bay's most productive winter flounder areas. From its eastern mouth at the inlet, along the western side of Clam Island and all the way to its western mouth near the #68 marker, Double Creek sports a nice mud and muddy sand bottom, deep pockets, and contours which warrant investigation especially during the spring. From a flounder fishing perspective, fishing is typically good at Double Creek because it has the bottom type flounder love to bury themselves in to escape the cold, and it flows adjacent to eel grass flats, which tend to warm up sooner than deeper surrounding water on sunny spring days.

Anglers can basically find winter flounder along Double Creek's entire stretch. Early in the season, during the late fall and early winter, look for flounder at the channel's eastern end. During the spring, flounder can be found from the #68 marker all the way east to the inlet with Clam Island usually being productive.

Some of the channel's better spots are the ones that attract fluke during the summer. For example, the junction point where Double Creek and a small second channel that heads south along Long Beach Island is generally productive. Another location worth investigating is where the channel flows past Clam Island. Fishing along the edges of the channel or even dead center, when there is no threat of boat traffic because anchoring in the channel is illegal, are where the flounder catches are put together.

One item to consider when fishing Double Creek during the spring is some years the channel lays unmarked until the summer. Meaning, anglers either have to travel north to Oyster Creek and then snake around to Double Creek or stick to the deep water near the marsh and then follow the deep water off the sand bar. The boat armada that usually anchors in Double Creek by Clam Island will point you to the channel. Or, employ a depth finder and some patience and navigate the unmarked channel. Regardless of whether the state marks Double Creek, the thick armada of boats that typically crowd the channel near Clam Island sort of gives one the idea of where Double Creek is located and good fishing is to be had.

One other favorite flounder area is the water between Double Creek and the #68 marker. Though the bottom here is generally flat, anglers who chum heavy and fish down current from the area's clam beds or near small, deep pockets pick up a healthy share.

Looking for flounder areas south of Double Creek, many anglers explore the Gulf Point area. One spot to check out in the Gulf Point area, is the channel that runs east of the Point, past the Sloop Sedge and south along Long Beach Island. Here, anglers chum the channel's deep pockets. Following this channel eastward, anglers will discover a hole that drops past the 15-foot mark. Here, anglers typically have productive days during the spring with the fall being not as productive.

Chapter 7
Blowfish Etcetera

In the fervor to catch the previously discussed species, many anglers fishing Barnegat Bay never fish for many other species that populate the bay. Though species such as: blowfish, kingfish, porgy, Spanish mackerel, and eel are usually an unintended "by-catch," some anglers, particularly younger ones, prefer to solely target these species. Why not? Besides being plentiful and fun to catch, many of these fish, particularly blowfish and Spanish mackerel, rival, if not surpass, the table qualities of the more "revered" species.

Blowfish

If New Jersey saltwater fishermen were to rank local saltwater fish species according to overall popularity, the blowfish would occupy the lower rung if it even managed to somehow make the list. If, however, these same anglers only considered table qualities during polling, and had the benefit of tasting a blowfish before voting, I am confident that blowfish would either top the list or be a runner-up.

The blowfish derives its name from its unique ability to inflate its body with air or water as a defense against its predators. Also known as "northern puffer," "puffer," "swellfish," or "ballonfish," blowfish (Sphoeroides maculatus) sport a scaless skin covered with short bristles. These bristles give the blowfish's skin a rough, sandpapery-like texture. Besides their tough skin, blowfish possess a small hard-boned mouth which contains two prominent, woodchuck-like teeth that are actually curved bone rather than true teeth. Blowfish use their "teeth" to easily nip and consume their shell clad prey. As stated earlier, when a blowfish perceives danger, it will inflate itself. The blowfish can significantly increase its size and with its rough skin, becomes quite unappetizing to a large predator fish.

Blowfish are colorful fish, possessing a dark brown to olive green upper body which fades into yellow or orange shading that then fades into a bright white along the blowfish's stomach. Distinct black bars line the

A smiling angler with a large Barnegat Bay blowfish that could not resist a thin squid strip. Note the stiff bristles concentrated on the fish's underbelly (opposite).

blowfish's flanks from head to tail while its fins are small and luminescent.

Physically, blowfish do not reach tackle-busting proportions, but they can, according to the New Jersey State record set in 1987, grow to a surprising 1-pound, 14-ounces. Within Barnegat Bay, blowfish average around 6-inches long but quite a few larger, adult fish will run about 10-inches long. Adult blowfish are not the tennis ball-sized specimens that one observes in some joker's seine net puffing up amid gasps and finger pointing. Rather, they can reach over a foot in length, and a doubleheader of two such specimens

From an angling perspective, blowfish have a voracious appetite and will aggressively attack all the previously mentioned organisms if used as bait in addition to squid strips and even live killies. Once while drifting for fluke, I slowly retrieved my killie from the bottom to check for weeds. As the killie approached the surface, I noticed an odd shaped blobby-looking fish following the killie. The fish, which turned out to be a large blowfish, quickly approached the killie and then stopped right behind the baitfish. With one split second lunge, the blowfish nipped the killie's entire body and then

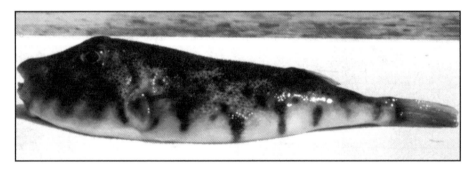

would surely give light tackle anglers something to tango with.

Blowfish Baits

A blowfish's diet primarily consists of shell clad organisms. In the marine environment, crabs, shrimp, mussels, barnacles, clams, snails, sea urchins, and sea worms find their way into a blowfish's stomach. If you ever looked at a blowfish's stout, strong buck "teeth," one can understand how they make short work of organisms, particularly crabs, that rely on their shells as a primary defense.

slowly sauntered below the surface. Only the killie's head remained.

Clam and squid strips make the best and most economical baits for blowfish. For clam strips, one should cut strips from the clam's tough tongue. Hardshell, steamer and surf clams all work for blowfish, but anglers primarily use surf clams because they can get more bait from these larger size clams. For squid strips, use the same type strip you would for fluke or weakfish except tone it down to compensate for a blowfish's small mouth. Strips should be cut so they are short and thin, roughly about 2-inches long and $1/4$-inchwide. When fishing clam or squid strips, be

A deflated blowfish. Note the fish's small beak-like mouth, that it uses to easily crush crustaceans and nip baits. The tough bristles located primarily on the fish's back and stomach may be uncomfortable to some anglers when handling a blowfish (center).

sure to double hook your strip bait so a blowfish does not pull it off with one nip.

Though bloodworms and sandworms make excellent blowfish bait, I am sure even the biggest blowfish fan is not in love with blowfish enough to bait his or her hook with these expensive baits. Their retail price, combined with a hungry blowfish horde, makes using sea worms a viable option only if you have them left over from a another fishing trip, you can dig them yourself or you have stock in the bait retail industry. When using a sea worm, be sure to thread it up your hook's shank, trying to avoid leaving a long piece dangling off. Blowfish, similar to porgies, will quickly tear off a worm once they get their mouths on it.

Tackle & Techniques

Blowfishing is a light tackle affair. A 5½ to 7-foot stiff-tipped graphite model rated for 6 to 15-pound line, equipped with either a conventional or spinning reel loaded with about 10-pound test is good for blowfishing. You don't want to go too light where you put yourself at a disadvantage if a weakfish or a bluefish grabs your line, but want to have the muscle to land a doubleheader of large blowfish.

A good rig for blowfishing starts by snelling a Mustad hollow point beak long shank hook to a 20-inch piece of 15-pound leader material. Snell a second hook to a 10-inch piece of leader and tie surgeon end loops at the end of both leaders.

Anglers typically employ hi-lo rigs or fish-finder rigs when blowfishing. The hi-lo rig is more widely used for blowfish because it allows anglers to work two different depths and employs minimal hardware. This rig is perfect when chumming blowfish from an anchored boat or when fishing from a stationary position, such as a pier. The fish-finder style rig is good to employ when searching for blowfish. Blowfish often take the bait so light, because their sharp teeth can easily sever a strip bait and swim away undetected. Large blowfish simply inhale a baited hook, fleece it and then spit the hook out. The sensitivity one receives from a fish-finder rig makes detecting a blowfish much easier.

Though opinions about proficient hooks for blowfishing may vary, I prefer a Mustad hollow point beak long shank hook. Some tie blowfish rigs with long shanked Chestertowns, in sizes 8 or 9, or Carlisle hooks #6 through #8. Whatever type hook you choose, make sure it possesses a thin gap, so a blowfish can easily fit it into its mouth, and a long shank so you can easily remove the hook. Long shanked hooks are a smart choice for blowfishing because blowfish, particularly large ones, will inhale a short shanked hook, such as a small baitholder model. You will spend more time unhooking blowfish and tying on new hooks than fishing. It's also smart to use a long shanked hook in case an oyster cracker takes an interest in your bait . . . .

After snelling the hooks to the leader, tie a drooper loop about 5-inches before the hook on the 15-inch leader, and attach the 10-inch leaderd hook to the loop. You then connect this rig to either a fish-finder rig or the bottom of a hi-lo rig. When fishing for blowfish, remember to hook your bait twice so it is firmly anchored on the hook so a blowfish is not able to knock the strip off in one fell swoop.

Many anglers think blowfish take the bait akin to a sea robin or small sea bass — a flurry of quick nips. Small blowfish tend to follow this pattern but larger cantaloupe-sized specimens usually take the bait similar to a crab or small fluke — a slight increase of weight. Unlike a crab or fluke, the blowfish's buckteeth will make short work of your bait. One way to figure out if blowfish are attacking a strip bait is to inspect the strip for the telltale sign of a blowfish: a steady stream of thick horizontal bite marks running the strip's length. If a strip possesses these type bite markings, re-bait with a shorter strip. Now that you know blowfish are in the neighborhood, bounce your sinker so it just barely touches the bottom. Wait until you feel a slight increase in weight or a peck and set the hook with a quick snap of the wrist. The result will usually be a blowfish, that will puff up when introduced to you.

Chumming

Once you have located a blowfish school, you can make repeated drifts over it but when you consider the current, wind and boat traffic it is often a lot easier to anchor up and chum the fish to the boat. Chumming is a leisurely but effective way to catch a bunch of blowfish.

When considering chumming for blowfish, the first thing to do is select an area to chum. Good areas to chum are any locations that attract and hold blowfish which include: channel edges, bridges, rough bottom, clam beds, mussel beds, and coral beds. Make a few drifts over a potential area to locate where the blowfish seem to be concentrated. Once you figure this out, anchor the boat up current from the location via the same double anchoring technique employed for winter flounder fishing. If the winds and boat traffic are light, try single anchoring. When trying to anchor over rough bottom, employ a long lead and anchor chain if conditions permit because it can sometimes be difficult to get the anchor caught in an unyielding type bottom. Once anchored up, you should not have a problem attracting blowfish via chum if chumming in a relatively small area, such as a clam bed, but keeping a large school around and interested is another story!

When it comes to chum, blowfish respond very favorably to the same type chums employed for winter flounder. The typical tackle store clam or mussel chum logs will do the trick with clam chum edging out mussel chum as the best choice. Bunker chum tends to attract legions of snappers. If you are one of those "do-it-yourself" types, a few shucked clams, pulverized mussels or "crab leftovers" will also attract blowfish. The leftovers from a crab cookout (i.e. entrails and shells) will greatly enhance a chum slick provided they are somewhat fresh and not rancid. Cracking whole clams or mussels and allowing the crushed shells to emanate their scent near the boat makes a good chum slick especially if you try this technique near a mussel encrusted structure or clam bed.

Whatever form your chum

takes, be sure to keep it near the bottom, similar to winter flounder fishing, because blowfish tend to hang near the bottom. If you chum too high off the bottom, you will probably spend most of the day fighting off hordes of snappers rather than blowfish. Make sure to bring enough chum because a swift current will quickly disperse finely ground chum. Once a chum slick weakens, blowfish will slowly disperse.

Cleaning Up

Now that you have caught a few blowfish you may be asking "How do I clean them?" Cleaning blowfish is hardly the heinous job many anglers make it out to be, provided you employ a long, sharp fillet knife, pliers and some know-how.

Initially, you want to sharpen your knife. Do a thorough sharpening job because a dull knife can complicate the relatively easy job of cutting through the blowfish's tough skin. You may want to wear gloves for protection from the minute bristles that cover most of the blowfish's body. These can sometimes irritate hands.

Make your first cut slightly above the blowfish's dorsal fin, where you feel the firm tail muscle fade into the stomach region. Cut straight through the fish, severing the tail from the body. After you cut the tail completely off, you should be able to see, looking at the tail straight on, the backbone flanked by solid meat. Wipe any stray innards, which are toxic, away from the meat.

The next order of business is to sever the dorsal and anal fins from the actual meat. Slide your knife under the skin and delicately cut these two fins so they are no longer connected to the tail muscle. Once you accomplish the "fin severing," take your pliers and pull the skin down. If you did a good job cutting the fins, the skin should almost slide off by itself.

By now, you should have the entire pale-colored tail muscle exposed. Take your knife and slice parallel to the backbone, taking the meat off from both sides of the backbone. You should now have two "blowfish fingers."

Whether you deep fry or broil them, "blowfish fingers" will rival the best seafood you have ever experienced.

Where and When

Blowfish range from Cape Cod, Massachusetts to Florida's Eastern Coast. They are bottom dwellers that prefer clam beds, mussel encrusted structures, rough bottom, eel grass flats, and coral beds. All these type habitats host the crustaceans, amphipods and mollusks blowfish prey upon.

Sometime around late April or early May, blowfish invade Barnegat Bay eating almost anything that comes their way — much to the chagrin of winter flounder anglers. As May wears on, blowfish numbers usually steadily increase and fishing for them is favorable into June. Around mid to late June, however, blowfish almost disappear from the fishing scene except for when anglers happen upon an occasional school while drifting for fluke. Blowfishing is steady provided one fishes areas that blowfish frequent, such as eel grass flats. Around mid-August, blowfish become apparent once again, providing great fishing until late

Did You Know?

The powerful toxin Tetraodontidae is concentrated in a blowfish's skin, liver and roe. Though one study found a person would have to consume 600 blowfish tails in a single sitting to feel any adverse effects, one should stick to eating just the tail meat and avoid eating any other parts.

September. By mid-October, anglers can usually find a smattering of blowfish but the bulk of them have bid farewell until next spring.

Within the Bay

In the upper bay area look for blowfish in the waters around Bay Head. Anglers can usually find blowfish near Shoals, the clam beds that lie southeast of Cedar Creek host a healthy blowfish population.

Popular crabbing areas that lie east and south of Cedar Creek are usually good blowfish spots simply because blowfish like to eat the crabs, shrimp and other crustaceans the area attracts.

depressions and within the channel that winds from the Point Pleasant Canal towards Herring Island. Check around the channel's buoy chains for blowfish because they house the small organisms blowfish like to feed upon.

Southward, the Mantoloking Bridge, the Route 37 Bridge, the Toms River and Tices Shoals all hold blowfish. At the Mantoloking and Route 37 Bridges, check near the bridges' foundations. At the Toms River, many of the areas that are good for crabbing are also good for blowfish, such as Goodluck Point. Over at Tices Shoals, explore the deeper water off the shoals to the west.

Across the bay from Tices

If fishing in the lower bay, look in Oyster Creek by the power plant discharge, Oyster Creek Channel, the Mud Channel, and Double Creek Channel. The channels are your best bet for a summer long shot at blowfish.

Perhaps a blowfish's favorite haunt is the coral beds that dot parts of the bay. Though I highly doubt the creature I am alluding to is actually coral, I will refer to it as "coral" for lack of a better word. The creature in question closely resembles the white tubular organism that grows on clam and oyster shells. However, this creature grows together in large clumps rather than on a shell. If you ever happen to examine a piece of the "blowfish coral"

Blowfish, kingfish and a host of other structure-oriented fish favor the coral-like beds that dot a few areas within Barnegat Bay. The coral that comprise these beds house a variety of marine crustaceans, arthropods and isopods that blowfish prey upon (center).

you will see a myriad of crustaceans and marine arthropods attached to, or hiding, in the coral's nooks and crannies. Not surprisingly, whenever I clean blowfish, I usually find pieces of this coral in their stomachs. Locating coral beds is essential if you want to consistently catch blowfish from spring to autumn because blowfish stay close to coral beds similar to the way tog hang

Kingfish

Every year, Barnegat Bay hosts a sizable northern kingfish population. A member of the croaker family, the northern kingfish (Menticirrhus saxatilia), more commonly known as "kingfish," is a favorite among many anglers who enjoy its fine table qualities

close to a rock pile or a wreck. You will know when you are over a coral bed by noticing how your bait receives a barrage of hits whenever you drift over a certain area. Coral beds attract kingfish, oyster crackers, porgies, sea bass, snappers and even weakfish. Be prepared to take an oyster cracker or two off the hook when fishing around coral beds!

Besides the channels, anglers can also find blowfish off Gulf Point and along the channel that winds east towards Long Beach Island. They are one of the bay's more aggressive species so if you happen upon a school, blowfish will usually make their presence known.

plus its relatively large size.

Kingfish are identifiable by a single, short fleshy barbel on their chin and a small, stout turned down mouth that allows them to easily scour the bottom for food. Dark gray bars mark their light gray, thickly scaled body which camouflages the kingfish against the soft-toned, sandy bottoms they prefer to inhabit. The kingfish's large, dark, sail-like dorsal fin easily distinguishes the kingfish from other fish that swim the bay.

Though most kingfish average less than a pound, they can attain weights up to 6-pounds. New Jersey's current record for kingfish, set at Barnegat Inlet, stands at 2-pounds, 3-ounces.

The kingfish is an interesting looking fish. Its general appearance resembles a croaker. Note the kingfish's small, turned-down mouth that the fish uses to quickly suck up baits while avoiding getting hooked (center).

Food Fit for a King

When it comes to bait, kingfish and blowfish share many of the same tastes. Kingfish respond favorably to squid strips, shedder crab, mussels, clams, bloodworms, and sandworms but with kingfish, the latter three baits are the best choices.

Many anglers consider clams to be the best kingfish bait. Anglers chum with cracked clams or shredded clams and then employ these mollusks as bait to catch the kingfish they lure. Surf clams generally make the best bait, but kingfish will also respond favorably to hardshells or steamers. Small, slender strips cut from a surf clam's tongue so a small bit of the clam's belly or ligaments hang off work great for kings. Fashion the strips blowfish-style, about 2-inches long and about $1/4$-inch wide to accommodate the kingfish's small mouth.

Bloodworms and sandworms in many anglers' opinions rival, if not surpass, clams as the best kingfish bait. When you have located a kingfish school, or are fishing a chum slick designed for kingfish, use 1 to $1^1/2$-inch sea worm pieces because longer pieces will get quickly knocked off by kingfish. While drifting for and trying to locate kingfish, employ a slightly longer worm, about 4-inches long, to grab a kingfish's attention. Similar to flounder fishing, use juicy worm pieces oozing with scent to produce the best catches.

A cheap alternative to sea worms that accounts for kingfish is squid strips. Small, blowfish-style squid strips really lend themselves to catching the small mouthed kingfish. Often while fishing for blowfish, I will catch a lot of kingfish not only because they share many of the same locations with blowfish, but also because they like squid strips. Dyeing squid yellow, so it appears clam-like, often produces kingfish. Squid is usually better to employ when fishing a chum slick, and when you have a captive audience.

Tackle & Techniques

The same light tackle outfit I described earlier for blowfish is perfect for catching kingfish. The two popular rigs for kingfish are the standard fish-finder rig and a modified drooper loop rig.

On the drift, the best rig to employ for kingfish is the fish-finder rig which lets the angler feel and react quickly to a kingfish's rapid, peck-like bites. The fish-finder rig for kingfishing is nearly identical to the one aimed at fluke except employ a hollow point long shank blowfish style hook or a winter flounder style Sprout hook so you can easily hook these small mouthed fish. Try a 15 to 24-inch leader rather than a 3-foot leader.

The drooper loop rig is basically a modified version of the drooper loop winter flounder rig. This rig is very good to employ when chumming for kingfish from an anchored boat. Tie this rig the same as you would for winter flounder, the only difference is use heavier leader. Try 10 or 15-pound leader to deal with the weakfish which often prowl the same areas as kingfish.

Though most kingfish caught by bay anglers are the result of a "by-catch," anglers who locate kingfish holding areas and then employ chum for the rascals catch more kingfish than many anglers think exist in the bay. Anglers can chum for kingfish with the same clam chum logs and weighted chum pots used for blowfish, or they can try a much more productive avenue which centers around cracked clams.

Just like winter flounder anglers do, experienced kingfish anglers locate a potential kingfish area, anchor upcurrent from it and then toss cracked clams near their boat. They usually start the slick with about five clams, and let the sweet clam juice draw kingfish to their boat. They then shuck a clam and bait up with a fresh strip cut from the clam's tough tongue. When it's apparent that kingfish are in the slick and are biting, they minimize the clams to about one or two. They want to keep the kingfish interested but not fed by the chum. Once their biting becomes infrequent, drop about three more cracked clams near the boat to rile the kingfish up.

When fishing for kingfish, be quick to strike when you feel a kingfish's unique jackhammer-like bites. Part of the reason why kingfish are not frequently caught is not because these fish don't give anglers any opportunities. It's rather because anglers do not strike back fast enough. If you give a kingfish line, like you are fishing for fluke, or pause for a solid hit, like when togfishing, kingfish will usually clean your hook before your strike comes. If, on the other hand, you respond to their bites quickly, like you are fishing for blowfish, then you will increase your kingfish score significantly.

Where and When

Kingfish occur from Maine to Florida where they prevail near sand bars, channel edges, clam beds, sandy shoals, river mouths, and creek mouths. They prefer sand, shell and gravel bottoms with contours and relatively deep cuts from where they will usually hang down tide to ambush their prey. Anglers operating along the surf, typically enjoy solid kingfishing throughout the summer.

At Barnegat Bay, anglers start catching kingfish around mid to late June. The fishing steadily picks up and usually reaches its zenith around late August. From then to mid-September, kingfish become much more conspicuous in the bay. By mid-October, kingfishing typically slacks off significantly enough that anglers turn their attention to other species. However, early bird winter flounder anglers occasionally find kingfish in their slicks as late as Halloween during some years. Be sure to have a net on hand because the bay's kingfish can get quite large. It was only a few years ago that the state record kingfish was swimming around Barnegat Bay.

The King's Court

Anglers looking for kingfish around the upper bay should investigate the channel the flows past Herring Island's eastern shores. Kingfishing is usually productive here during the late summer for anglers who take time to locate the channel's slight contours and pockets kingfish favor.

Another notable place in the upper bay for kingfish is the Mantoloking Bridge. Anglers should check within close proximity to the bridge for the better fishing. During the late summer, anglers poking around the channel that winds beneath the bridge will pick up kingfish.

Moving closer to the lower bay, Tices Shoals is one particular location that plays host to a healthy kingfish population. Though most kingfish caught here are only small 8-inch rascals that go after sandworms intended for weakfish, there are a few nice size fish that occasionally grace the scene. If you ever try chumming for weakfish at the shoals, drop a few crushed clams off the boat if things are not going well and explore this location's great kingfish possibilities.

Within the Oyster Creek Channel area, kingfish tend to stay concentrated at the channel's eastern mouth, especially along the channel's sandy bottomed parts. Anglers who take the short jaunt to the Mud Channel catch kingfish usually during the late summer.

Arguably the best kingfishing in the bay is to be had at Barnegat Inlet. Anglers should especially investigate the inlet's western mouth, where it meets Oyster Creek and Double Creek Channels. Here the bottom is sandy and deep, and anglers fishing with clams and sandworms typically put together the kingfish catches. Chumming here is tricky because of the constant boat barrage and the inlet's strong currents. Try chumming early or late during a slack tide to have a good shot at the kingfish.

A short hop from the inlet, Double Creek Channel's eastern mouth is a very good place to catch kingfish during the late summer. Anglers can also find kingfish where Double Creek flows past Clam Island usually keeping company with small sea bass and blowfish.

Porgy

The porgy is usually an incidental catch of either a tautog angler fishing a rock pile or a blowfish angler fishing a clam or coral bed.

Also known as "scup," the porgy (Stenotomus chrysops) is a rather attractive fish that resembles a freshwater sunfish in general appearance. It is a thickly scaled fish that possesses large spines along its dorsal fin and ranges in color from a bright, pure silver to a dull, tarnished silver. Porgies often have a reddish hue running through their brilliant silver coloration.

Though porgies can reach 18-inches in length, these large, "plate-sized" fish are generally found at offshore wrecks. Many anglers target these large porgies because they make excellent table fare. Within Barnegat Bay, most porgies take the form of small 5-inch bait stealing rascals but there are a few 8-inchers occasionally mixed in with the diminutive populous.

The Porgy Palate

Porgies respond to both bait and lures. Clams, squid, mussels, shedder crabs, sea worms, or shrimp are standard fare. Small clam strips, squid strips and sea worm pieces make the best overall baits.

Small strips cut from a surf clam or squid make great porgy bait. Strips should be cut about 2-inches long and ¼-inch wide, but if you find porgies still tearing your bait off quickly, use shorter strips.

When using sea worms, which porgies absolutely love, anglers should use small, 1-inch pieces to avoid having a porgy knock off the bait with one hit. Anglers can employ a 2-inch piece but be sure to thread the worm piece your rig's hook. Varying your rod and reel to a light setup is not advisable in these situations because the larger species you were originally targeting may tear into the bait attached to a light tackle outfit and will be hard to land.

When fishing for porgies, fish-finder rigs are good when drifting over porgy areas, while a hi-lo rig or modified tautog rig is great when fishing from a stationary position. The previously described blowfish rig will

carefully up the hook's shank. The same goes for strip baits and other soft baits; thread the bait carefully onto your hook so not much hangs off. Letting a large part of the bait dangle off the hook is an invitation for a steal.

The best tackle for porgy fishing is the same light tackle gear employed for blowfish and kingfish. If anything, be sure to employ a stiff-tipped rod so you can feel a porgy's small nips. Overly flexible rods make it difficult to feel porgies.

If you initially target another species, such as tautog, and then switch over to porgies, you need to only alter also account for porgies.

When tying a rig for porgies, employ small hooks so you can easily catch these nibblers. The most widely chosen porgy hooks are Chestertowns, typically size 8 or 9, a #7 Virginia or the Mustad hollow point blowfish hook mentioned earlier. All of these hooks possess a narrow gap, so a porgy can easily fit part of the hook in its mouth while the hooks' long shank makes for easy hook removal.

Similar to kingfishing, you have to be quick with the trigger when porgy fishing. When you feel a porgy's distinctive quick, successive "pecks,"

A small mouth coupled with a voracious appetite makes the porgy a proficient bait-stealer. Note the fish's erect spiny dorsal fin that can easily cut hands. Use a towel when taking one of these characters off the hook (center).

strike back immediately or the porgy will strip your hook.

Where and When

Porgies are found from Nova Scotia, Canada to Florida's Eastern Coast. Similar to sea bass, porgies are bottom dwellers that prefer rough or gravel bottoms, clam beds, wreck sites, rocks piles, pilings and most other structures natural or man-made. Anglers will often find porgies accompanying blowfish, kingfish, sea bass, and tog.

Though anglers can catch porgies on inshore wrecks during the late spring, anglers fishing within the bay begin to encounter porgies around July. August and early September generally provide productive fishing if anglers target porgy-friendly areas, such as a rock pile. During September and early October, anglers can usually find porgies populating the inshore wrecks that lie off Island Beach and Long Beach Island. If you do not mind the cold, take a shot at the large, "plate-sized" porgies which are usual catches aboard offshore winter wreck trips.

The two prime porgy attracting areas within the upper bay, save for the Point Pleasant Canal itself, are the Mantoloking Bridge and the Route 37 Bridge. When investigating these two locations, look for porgies to be holding very close to either bridge's structure. Carefully aiming a hi-lo rig baited with bloodworms or clam strips close to the bridge's structure will often yield porgies.

In the lower bay, look for porgies holding near the clam beds that lie off Cedar Creek and the coral beds that dot the bay. Anglers chumming within the channels kingfish-style will typically catch small, "silver dollar" porgies and maybe a few decent sized specimens.

Barnegat Inlet is another porgy holding location. When looking for porgies, check the inlet's rock piles, particularly its north and south jetties, and where it wraps around Barnegat Lighthouse. The pockets where one would find tautog often hold porgies. In the lower bay, look for porgies during the mid-August to early September time frame.

Spanish Mackerel

Distinguished with a bright silvery body emblazoned with several large bright green spots, the Spanish mackerel is a warm water species that frequents New Jersey's waters when the ocean temperature explodes into the mid-seventies during the height of summer. Though these tropical visitors are mostly ocean dwellers, they will follow baitfish schools into Barnegat Inlet, giving anglers a nice surprise.

Resembling Atlantic and tinker mackerel in general appearance, a Spanish mackerel (Scomberomorus maculatus) is easily distinguishable from these other mackerels by its bright, prominent greenish yellow spots that dot its chrome-like flanks. A stiff, tuna-like tail gives the Spanish mackerel lighting speed and its strong, tooth-filled mouth makes the fish very adept at tearing through a baitfish school.

Did You Know?

The porgy was one of the fish species to take a huge hit by foreign fishing fleets during the 1960s.

The largest Spanish mackerel ever landed on rod and reel was 13-pounds and was taken off North Carolina. The New Jersey record, set in 1990 off Cape May, weighed 9-pounds, 12-ounces. The more common size for the Barnegat Bay area is about 5-pounds, though larger sized fish will make an appearance.

If debating about whether or not to eat a Spanish mackerel, do not let the "mackerel" moniker fool you into thinking this fish is oily fleshed, better used for crabs bait. Part of a Spanish mackerel's appeal to anglers, besides its strength is its delicious table qualities. Try one of these firm, white fleshed fish baked or broiled, and you'll be hooked.

Tackle & Techniques

Fishing for Spanish mackerel generally consists of spotting surfacing fish and either trolling lures by the school or casting at the school with small lures. Because the Spaniards usually position themselves at the inlet 's tumultuous parts, namely where it meets the Atlantic, chumming from an anchored boat is usually out of the question.

Spotcasting for Spanish mackerel, similar to bluefishing, consists of spotting breaking fish, which sea gulls usually pinpoint, and retrieving lures through the breaking fish. Be careful when approaching Spanish mackerel breaking the surface. Though they can tolerate a few passing boats, the dummies whose course curiously cuts through the school's center will usually disperse the fish or send them below the surface.

Though Spanish mackerel will often attack with the same reckless abandon and ferocity as bluefish, you cannot just throw any old lure and hope to catch a Spaniard. Lures that match

the silverly, diminutive baitfish the spaniards love to feed on, such as spearing and rainfish, are the key components to a successful outing whether you plan on spotcasting or trolling. Quarter-ounce Bridgeport jigs, 1/4-ounce Hopkins lures, and small Kastmasters are good lures for Spanish mackerel when worked at "warp speed" so they skip and dart along the surface to imitate frantic, fleeing baitfish. Small,

Luke Henderson and Chelsea pose with a nice Spaniard that fell to a trolled Clark spoon. Anglers should make it a point to investigate an apparent bluefish school during the summer because some schools are either mixed with or composed entirely of Spanish mackerel (right center).

¼-ounce pure white bucktails worked quickly along the surface will also account for Spanish mackerel.

When casting lures for Spanish mackerel, make sure to tie your lure directly to your line. Spaniards have very keen eyesight and can easily detect a wire leader, no matter how fine, especially in the inlet's clear waters. Although they possess sharp, bluefish-like teeth, leave the wire leader in the tackle box. Some anglers employ a foot long piece of shock leader to compensate for the lack of wire leader. If you plan to go the shock leader route, employ clear line to better blend with the water.

A 6 to 7½-foot, stiff-tipped graphite rod identical to the one one would use for bluefish is a good rod for working lures for mackerel. Combining this rod with a spinning reel, with a gear ration of 6:1 loaded with 10-pound test, is a good outfit for: casting small metals to Spanish mackerel, retrieving the lures at warp speed along the surface and fighting the fish. Anglers basically use the same medium spinning tackle they do for bluefish. Though the Spanish mackerel angler is likely to catch runs larger than a tailor bluefish, the somewhat light tackle can handle the mackerel and make for some exciting fishing.

Trolling typically comes into play when anglers target mackerel that have fallen below the surface. Spanish mackerel will typically swim deeper once the water's surface becomes too bright or when too much boat traffic sends a baitfish school down deep. Many anglers favor trolling over spot casting because they can keep their boat moving rather than stopping it to cast in the often tumultuous inlet.

Trolling for Spanish mackerel is relatively easy; it usually consists of tying a lure to a rod and dragging the lure either slightly below the surface or further down in the water column via a moving boat. Five-and-one-half to 7-foot rods rated for 20 to 60-pound line, equipped with a conventional style reel loaded with 20-pound test are good outfits to troll Spaniards with. Your lure selections should include: ½ to 1½-ounce Jap feathers, in pure white, white/red or blue/red and 0 or 00 Clark Spoons. Tie one of these lures to a 10 to 15-foot piece of 20 to 30-pound line. Drop down to 15 or 10-pound leader if the water is clear, and you're not receiving any hits. Chances are the mackerel are not hitting your lures because they can see the leader. Connect the other end of the line to a ball-bearing swivel, such as a Sampo, to prevent the line from twisting during trolling.

Locate a school of breaking fish or try trolling an area with no apparent action. Many times, mackerel will hang below the surface and be undetectable.

To troll, drop the tied lure in the water while driving towards the area you wish to troll. Let out about 50-yards of line. Do not engage the reel but put its clicker on so a mackerel can easily take line when it grabs the lure. If aiming your boat at breaking fish, troll the outskirts of the breaking fish so as not to disturb them. Troll the lures along the surface at a reasonably brisk pace of about four or five knots. Use this speed as a guide because you will often have to play with the trolling speed when there's a strong current or wind. Once a fish grabs your line, which is signaled by line squealing from your reel, engage the reel and strike back.

Savvy mackerel hunters troll along the surface early in the morning or at dusk when its relatively dim and there is little or no boat traffic. When trolling Spanish mackerel at the inlet, you do not generally need to employ drails or planers to drop your lures deep because the inlet's eastern mouth is relatively shallow. Drails and planers are typically employed when exploring the inshore waters off Long Beach Island or Island Beach.

Where and When

Spanish mackerel range from Brazil to the Chesapeake Bay. A tropical fish that prefers water in the seventy degree neighborhood, Spanish mackerel travel north to New Jersey around mid-July.

During mid-July to around mid-September, Spanish mackerel frequent the waters where Barnegat Inlet meets the Atlantic. A Spanish mackerel school's stay in Barnegat Bay is usually a brief one. Typically, they will enter Barnegat Inlet from the ocean on an incoming tide, ravage baitfish schools from the inlet's eastern mouth west to the lighthouse and then will fall back to the ocean on the outgoing tide. There, the Spaniards will feed on the baitfish that get flushed out of the bay.

These occurences primarily take place at dusk or early in the morning, with many anglers mistaking a Spanish mackerel school for bluefish. It pays to to probe an "apparent" bluefish school with small metal lures with a quick retrieve. Once the sun comes up

and the inlet's daily marine motorcade begins, anglers should look for Spanish mackerel prowling the inshore waters off Long Beach Island and Island Beach.

American Eel

The question: "What is that a snake?" usually greets an angler after he or she lands an American eel. When it comes to species that are looked at with both amazement and disgust, the American eel tops the list of many inside and outside the angling community. Often overlooked by the angling community, except as bait for striped bass, the American eel provides great sport for many anglers and excellent meals.

Also known as "common eel," "eel," or "snake," the American eel (Anguilla rostrata) possesses the classic serpentine eel physique and ranges in color from a dark greenish black to a light hunter green fading to a lighter colored underside. Eels possess small, Velcro-like teeth that allow them to firmly grasp their prey. From head to tail, eels sport a thick coating of slime that frustrates the most dexterous anglers trying to grasp them and causes disgust among many outside the angling community.

Eels can become quite large which is evidenced by one eel taken commercially that measured 48-inches in length and weighed 10-pounds, 15-ounces. This eel possessed an amazing 13½-inch girth. The all-tackle record eel weighed 9-pounds, 4-ounces and was taken at Cape May during 1995

Did You Know?

Eels are born in saltwater and then reside in freshwater or brackish water. After living as long as 20 years, they return to saltwater to spawn and die.

(Who says record size fish don't swim New Jersey's waters). Thus, anglers whose sole impression of eels are the foot long specimens heaved for stripers are in for a rude awakening. Try

A young writer lifts his "record" eel. This large specimen was taken from one of the many brackish lagoons carved into the bay's western shores (left center).

hooking into a 3-foot long "snake" on a cane pole!

The Eel Deal

Fortunately, eels are not the most finicky creature to ever grace the Earth. You do not need to scour the bait department for a particular bait because eels will eat a variety of baits both live and dead. Squid strips, clam chunks, fresh sand eels, live killies, nightcrawlers, and sea worms work great for eels. Small 3 to 4-inch clam strips, cut from a hardshell clam, or squid strips, will tempt an eel's palate. Eels also have quite a taste for bloodworms and small live or dead killies.

The best tackle to catch eels with is light, but offers the necessary muscle to bring in an anaconda-like character. Most experienced eel anglers use 5 to 6½-foot graphite spinning rods, loaded with 8 to 10-pound test. My personal favorite eel fishing rod is an old fashioned 7-foot cane pole with about 10-feet of 8-pound test tied to the end. When fishing shallow locations, I simply wrap excess line around the cane pole's tip to get an exact length so my bait rests on the bottom and the line is taut.

A #10 Eagle Claw hook or #8 Sprout hook snelled to the main line with a bb-size split shot placed about 5 to 8-inches above the hook is all the terminal tackle one needs to catch eels.

When fishing for eels, let your line out until the splitshot hits bottom and reel up the excess line so your line is tight. Lay your rod down on the ground or dock and wait for an eel to come along. When an eel grabs the bait, regardless of whether the eel is a small "shoe-lace" or a large anaconda-like specimen, eels are not a shy fish; their strong, distinctive pulls are easily noticed. Grab the rod and set the hook with a flick of the wrist. Once you feel an eel on a rod, you will always be able to distinguish them from other fish by their easily identifiable straight up and down tugging motion. Eels usually start this up/down tugging motion as they near the surface. For all you "big-game" anglers, a fat 2-foot eel caught with an ultra light outfit or a cane pole is nothing to laugh at!

Where and When

Eels prefer Barnegat Bay's brackish water, swampy areas, rivers, and creeks which are typically found along the bay's western shores. Within these particular locations, eels seem to prefer deep, slow moving areas with a muddy bottom, especially if located at the mouth of a river or creek.

Anglers can typically catch eels between late April to early October. Mid- June to early September usually brackets the prime fishing period. Once the water temperature drops during the autumn, eels bury themselves in the mud of the swamps they inhabit and wait for next spring's arrival.

When looking for eels in Barnegat Bay, explore the bay's western banks, from Beaver Dam Creek south to Barnegat Township. A few areas that stick out are: Beaver Dam Creek, the Metedeconk River, the Toms River, Cedar Creek, Stouts Creek, the Forked River, and the swamps situated along the bay. The numerous residential lagoons carved into the bay's western banks, primarily the ones found from Barnegat Township to the Forked River are good places to check for eels.

Chapter 8
Blue Claw Crabs

With the Barnegat Bay area boasting the current state record for blue claw crabs, a crab that measured 8 $^3/_4$-inches from shell tip to shell tip, it's only fitting that I devote a word to the blue claw. Blue claws are plentiful within the Barnegat Bay, fun to catch and widely regarded as a delicacy, so one does not have to travel south to the famed crab restaurants of Maryland to enjoy the taste of blue claw crabs.

The blue claw crab (Callinectes sapidus), also known as the "blue crab," is named for the bright blue coloration found along its claws and sometimes along its legs. The shade of blue ranges from a bright azure to a dull blue-grey with the brighter blue being found in full grown males rather than in females or juvenile crabs. Save for its snow-white to ivory colored belly, the rest of the crab is a dark brown. The dark brown coloration offers the blue claw a perfect camouflage against the dark, muddy bottoms of the salt marshes and brackish estuaries they love to inhabit.

Crabbers can ascertain a blue claw's size by measuring the width of the crab's shell from shell tip to shell tip. As of 1997, the minimum size for blue claws is 4$^1/_2$-inches.

Unlike most other marine species, blue claws have particular names bestowed upon them to describe their particular life cycle or gender. One could probably fill a small dictionary with all the names crabbers have invented over the years.

"Hardshell" or "hard" crabs refer to blue claws that possess a typical hard shell, which is their main defense against predators. Once a hard crab sheds its shell to increase its size, it takes about four days for the crab's new, but soft, shell to harden.

"Peeler" or "shedder" crabs are hardshell crabs that are about to shed their shells so they can attain a greater size. Underneath a shedder crab's hard shell is a fully formed soft shell. Shedder crab is arguably the best bait for weakfish. Weakfish anglers peel shedders like an orange and then cut them into chunks.

"Buster" crabs refer to blue claws that have begun to emerge from their old shell. When molting, the crab's shell splits along its back crease where the flippers join the body, and the crab backs or "busts" out of its old shell. Blue

A bucketful of large blue claws is the aim of many people who ply Barnegat Bay's waters during the summer. The bay's blue claws are not the comatose specimens one finds in seafood markets or supermarkets (opposite).

claws cease eating when they are preparing to molt and seek a hideaway as protection from predators.

After emerging from its old shell, the blue claw, with its extremely soft shell, is called a "softshell" or

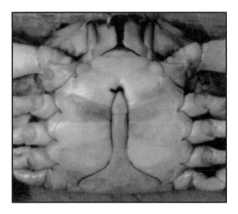

"softie." Softshell crabs are delicious to eat and also make good bait for a variety of fish species, particularly weakfish and fluke. While you will rarely catch softshells via a handline, because they prefer to stay hidden and not eat, you can catch them by scraping them off natural or man-made structures, usually at night. When you manage to catch a softshell, make sure to isolate it from any hardshell crabs you might have caught because the hardshells will tear their softshelled brethren apart.

Crabbers commonly refer to average-sized male crabs as "jimmies," while calling large, full grown males "channelers." Males grow much larger than their female counterparts and their coloration is much sharper and brighter. Also, the bottom part of a male's claw is blue rather than red.

Female blue claws, are differentiated from male crabs by having differently shaped abdominal aprons and red tipped claws rather than

the bright blue males sport. Crabbers commonly call juvenile female blue claws "she-crabs" and refer to mature females as "sooks."

When pregnant, a mature female's large, greyish apron opens and a large visible egg sac protrudes from the crab's underbelly. The crab's sponge-like egg sac, that is often as wide as the crab's abdomen and at least half as long, confers the name "sponge crab" upon the pregnant female. The egg sac will range in color depending upon the stage of the eggs' development. As the crab embryos grow and feed on the egg's yolk sac, the eggs will turn from yolky orange to a dull brown.

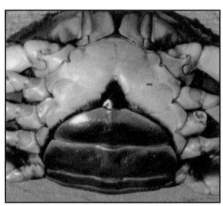

It's illegal to keep pregnant females, and crabbers should make all attempts to gently release these specimens. Ensuring a pregnant female's health is one important step to maintaining a healthy blue claw population and not killing a few hundred crabs in one fell swoop.

Catering To Crabs

The philosophy that preaches

Male blue claws, both juveniles and adults, are easily distguishable by the upside-down "T" shape of their abdominal apron (top left). Juvenile female blue claws possess a pyramid-shaped abdominal apron while mature females possess a semi-circular dark greyish apron (bottom right).

blue claw crabs will eat any old slop is practiced by those who do not consistently catch blue claws. Although having scavenger tendencies, blue claws prefer fresh food and have their preferences regarding what foods they will eat. One time while crabbing in a salt marsh, I watched as two crabs "herded" a spearing school against a marsh bank. The two crabs with their claws extended, positioned the fish against the bank so escape was quickly becoming an impossibility. Although the spearing opted to go over the crabs, this scene showed me that blue claws are not some lazy creature that lays on the bottom picking through the trash. Rather, they are aggressive hunters who use their quickness and sharp claws to catch their prey.

Good baits for blue claws are: chicken parts (typically the back or neck), bunker, mackerel, whole squid, killie rings, and filleted fluke or bluefish bodies. Blue claws tend to respond strongly to oily fish, such as bluefish, bunker and mackerel. When crabbing with oily fish, poke a few holes in the fish's side so its oils will leak into the water and attract blue claws. One thing to remember is not to use rancid bait. Baits with a scent are good but leaving them to "ferment" on a nice hot July day and then trying to crab with them a few days later is not a good idea — for crabs or crabbers!.

Baits made from fish that prey upon blue claws, such as blowfish or tautog, generally don't typically work great for blue claws. Supposedly, blue claws can recognize the bodies of their adversaries and shy away from them as if they were alive. All I know is, crabbing with such fish does not make for productive crabbing.

Ideal crab baits are fresh, but baits that have been frozen properly also work. If you securely wrap crab bait so it's air-tight before freezing, and then use the bait within a relatively short period of time, you're in business. When blue claws are thick during July, August and early September, they will surprise you with how aggressively they attack and hold onto a good bait even though your net or trap is closing in around them. Stick with fresh or freshly frozen baits, and you'll do well. If you try crabbing with bait that has been in the freezer since the '80's, then you are asking for, and probably will receive, an unproductive day.

Catchin' Crabs

Crabbers catch crabs through a variety of ways. Some of the more popular crabbing techniques found along Barnegat Bay involve: handlines, crab traps, scraping and scoop netting. While crabbers often disagree over the best method to catch crabs, one thing all crabbers will agree with is blue claws are tasty and well worth the effort.

Down To the Wire

Catching blue claw crabs with a handline is one of the true joys of summer. Handlines are popular with many crabbers because of their simplicity. A weighted bait attached to a line and a scoop net is all one needs. When crabbing with a handline, the bait is allowed to rest on the bottom, and when a crab seizes the bait, the crabber slowly retrieves the bait and nets the crab at the surface with a scoop net.

Did You Know?:

Callinectes, of the blue claw's scietific name Callinectes sapidus, mean "beautiful swimmer" while sapidus means "savory."

Did You Know?:
The largest blue claw crab was reportedly caught in the Cheseapeake bay and measured 9 fi-inches from shell tip to shell tip.

To construct a handline, all you need is aluminum wire, wire cutters, some twine or monofilament fishing line, a bank sinker, and a piece of wood or other flotation device. Initially, you want to cut about a 10-inch piece of aluminum wire. Stick with aluminum because it is very malleable and can easily twist to secure baits. When cutting the wire, 10-inches is an average size, but consider the size of the bait you are planning to crab with. You do not, for example, want to cut a 5-inch piece of wire if attempting to crab with a filleted fluke body. The short wire will make it hard to secure the bait so it can't resist getting ripped off by a big blue claw in an eye wink.

You then want to slide the wire through the eye of a bank sinker. A 2 or 3-ounce bank sinker is usually sufficient for crabbing around most areas in the bay. When crabbing near areas with stronger currents, such as Barnegat Inlet, you will have to employ heavier sinkers to keep the baits on the bottom. Once you center the sinker on the wire, cross the wire once over the sinker and then twist it a few times to firmly anchor the sinker into the centered position.

Finally, take some line and tie it to the bank sinker. Tie the line to the sinker rather than to the wire because the sinker will offer a solid connection while the wire will occasionally cause the bait to jiggle slightly as it moves. This slight jiggle may cause unwary crabs to dart off the bait. Attach a flotation device, such as a piece of wood, in case the handline drops in the water.

When considering line to employ for a handline, a smart and economical choice is old fishing line.

Not only do you recycle old line, but clear monofilament fishing line is likely less visible to blue claws than the standard tackle store bright white twine. Weight class of crabbing line varies from 12 to 20-pound test range. When considering employing 25-pound test or heavier line remember, heavier line will drag more in the water which will require more weight to keep the bait on the bottom. On the other hand, don't want to go ape and employ 6-pound test because such a thin line will bite into your hand when you attempt to retrieve a bait. After a few hours of crabbing, most will find that it would have been more comfortable to use heavier line. You then slide the line through the sinker's eye and attach the line to the sinker.

After tying the line to the sinker, you want to figure out how much line to use. For most areas in Barnegat Bay, 12 to 14-feet of line is sufficient. Though most good crabbing areas in the bay average around 6-feet deep, you want the extra line to compensate for the distance between the place you are tying the line from, usually a boat cleat or bulkhead, and the surface of the water. The water you plan to crab may be only 5-feet deep but it may be 2-feet from the water's surface to where you tied the line. Also, if you are crabbing from a boat, you want your bait to stay on the bottom. If you have the handline tight, so your bait is barely touching the bottom, and the Queen Mary blows by your position throwing up a sunami-like wake, your bait is going to move with the boat's swinging and bobbing. By employing 3 or 4-feet of extra line, you can quickly release an extra 2-feet of line and this excess line will significantly lessen a large wake's effect on your bait's position.

Perhaps the most important reason for using longer than average lines is because large blue claws often like to swim with the bait. If you try to horse them towards the surface, you will usually not get them up from the eyes, or in the case of a severed headless piece, run the wire centered on the body. For a filleted fish body, run one end of the wire through the fish's gills, out its mouth and then connect the wire's two ends.

After attaching the bait to the handline, drop the bait in the water, let

depths much less than get a clear shot at them with a net. A longer line allows you to follow the crab as it swims with the bait. Allowing a blue claw to swim with the bait will result in more large crabs and more triple or quadruple "headers."

Now that your handline is complete, attach bait to it by running one end of the wire through the bait and then twisting the wire's two ends together. Twist the wire a few times to make sure the bait is secure. For chicken parts, you can attach them by running the wire through the chicken's tough bony part. For bunker or mackerel, either run the wire through the fish's

it hit bottom, and then tie the line to a boat cleat or bulkhead. If crabbing from a boat, leave excess line so if the boat swings or bobs too much, the bait will still rest nicely on the bottom. Periodically check the line for crabs by slowly retrieving the bait. If you feel any change in weight or a crab pulling the bait, bring the line towards the surface very slowly. As mentioned earlier, if the crab swims left, follow it left. If it starts swimming away from your position, outstretch your arms so the crab or crabs eating the bait don't notice that they're being pulled towards the surface. Rather than hamper a crab's movements, move with them.

"Channlers" abound in Barnegat Bay's back water creeks and marshes. Keep a steady hand and quick net when trying to catch these large but quick characters (center).

Now here's the fun part: netting. A blue claw's typical escape route will be on a horizontal angle away from you. This means that if you want to net blue claws consistently, you need to position the net from behind the crab and swoop forward. The only thing you have to worry about is whether the crab will dart left or right. If, on the other hand, you choose to net a crab by scooping towards the crab's front, you will be testing your speed versus the crab's speed. You may succeed sometimes, but when large or multiple blue claws are involved, you will usually miss some of them this way because

similar to the way fish follow a chum slick. Rather than sitting around your position swatting at flies for 10 or 15 minutes as your bait defrosts in the bucket, give the bait a few pokes along its flank with a knife before dropping it to the bottom. Often, crabs will come exactly from where the current is taking the bait's juices towards.

Trapping

The white lines of crab traps tied to the bay's docks and piers are a common sight during the summer. Crab traps are popular among many crabbers because they do not require the

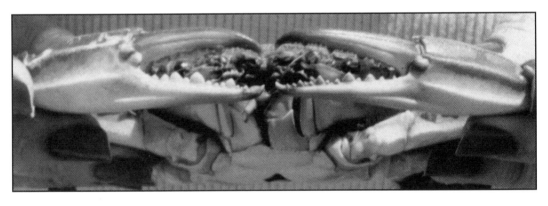

A blue claw's sharp claws allow it to easily tear apart baits or bloody a hand if one is not careful. Use a net or tongs when handling theses characters because blue claws may seem quiet and awkward on land, but are lightning quick with their claws. Take note wisenheimers (center)!

blue claws are extremely quick.

When netting, I would also recommend keeping the net above water prior to a net attempt. Blue claws are usually alert and will either see or sense a net placed in the water and take off. Also, water will place more resistance on the net than air, slowing your net down against something that is arguably faster than most crabbers' net to begin with.

When crabbing, remember to give a spot a chance. Provided you are employing a fresh or adequately frozen bait, crabs will follow the bait's scent

dexterity handlines need. With crab traps, crabbers simply attach a bait inside the trap, and blue claws enter the trap to get the bait. The small 4-door models and the large Maryland-style commercial traps are the more commonly employed crab traps along the bay.

With the small models, which typically come in either a cube or pyramid design, it's up to the crabber to trap the crab by quickly pulling the trap's line to close the trap. These type traps require more attention than the much larger commercial traps. When using small traps, give the trap a few practice drops on the ground or dock to see if it opens correctly. Many times, the trap's lines may be tangled causing the door not to open.

When throwing a trap in the water, you will want to make sure the trap rests on the bottom right side-up. If it's laying upside down or on its side, the trap is significantly hampered from catching crabs. A quick way to tell if the trap is laying right side up is to slightly pull its line. If resting on the bottom correctly, you should be able to feel the doors close. If you just feel a dead weight or actually feel the trap tumble along the bottom, chances are good that the trap is on its side and not positioned to catch crabs. Similar to using handlines, use more line than expected for a trap, to compensate for a boat's wake or deeper than expected areas and so it rests nicely on the bottom.

The other type crab trap commonly employed throughout the bay is the commercial Maryland-style trap. These traps, with their labyrinth-like interiors, entrap crabs themselves so all crabbers have to do is leisurely pick the traps up every few hours to catch several crabs at once. Commercial-style traps are sometimes as large as 3-feet by 3-feet and typically possess four to six openings from where crabs can easily enter the trap, to reach the bait at the trap's center, but can rarely find their way out of the trap's maze-like interior. Many crabbers fit these type traps with a long, sturdy line with an attached plastic jug. The baited trap is then dropped in a crab holding area and left with the jug marking the trap's location. If you choose to try this avenue, make sure the trap is properly rigged to the marker buoy because commercial traps if lost or left unchecked for days will kill crabs because most crabs will not be able to find their way out of the trap. They will starve and become cannibalistic. Note that there's a law requiring crabbers using commercial traps fitted with marker buoys to procure a permit. This permit is available at local tackle stores.

If you're going the crab trap route, you may want to invest in a plastic coated model. Besides resisting corrosion, coated crab traps are usually dark colored and blend more with the bottom. If you stick to the standard galvanized steel traps, letting the saltwater tarnish them a tad so they are a dull grey or light brown is good but don't go overboard because parts of the trap will simply rust out. Be sure to rinse uncoated traps with fresh water and give them an occasional light spray of WD-40. Even though the trap is made of galvanized steel, it will still rust out within a few weeks if neglected.

Scraping By

If you want to catch blue claws with the barest of necessity, you should try scrapping for them. Scraping, also known as "jacking," basically consists of spotting a blue claw attached to a piece of structure, which is usually a barnacle-encrusted piling or marshy embankment, and netting the crab off the structure with a quick scrape from a net. Whether you target old piers, bulkheadings, bridges or a salt marsh's banks, scrapping is generally better done at night or daybreak. In these low light conditions, you should bring a

Did You Know?

A lucky and healthy blue claw can expect to live 4 years.

Did You Know?
A blue claw may
molt 20 times
during its life,
increasing its width
¼ to 1/3 its previous
size every time it
molts.

powerful flashlight, so you can pinpoint an attached crab and an 8-foot plus net so you can easily net low lying crabs. On a good night, scrappers can often net more blue claws than if they waited until morning and broke out the handlines and crab traps. One thing to remember when perusing the bay at night, leave the Bermuda shorts and tank tops at home because the bay's resident mosquitoes and gnats are often vicious.

When scrapping structure at night, be sure to scan the nearby water's surface for free-swimming crabs. Locations that throw light onto the water, such as bridges, marinas and public docks, typically attract free-swimming crabs at night. On very still nights, one can often see the ripples caused from a large blue claw swimming along the surface — a beacon to a quick catch.

Take Care

After you have caught yourself a good bucketful of blue claws, it's important to keep them alive between the time you catch them and the time you cook them. Eating dead crabs presents certain health risks, so keeping blue claws alive is a top priority. Though blue claws are typically very hardy, there are four things you can do to safeguard their lives: keep them out of sun and water, keep them right-side up, and do not cram them.

Keeping blue claws cool and moist is key to keeping them alive, and you will not be able to accomplish this if a hot summer sun figures into the scene. Always keep blue claws shaded and never place them in a bucketful of

water. The sun will dry them out while if placed in water, they will deplete the oxygen level and drown. Your best bet is to place them in a shaded bucket. If need be, place a towel or board over the bucket allowing a space for air, to create your own shade. Occasionally, put some fresh saltwater on blue claws to refreshen them and then dump the water out. You will moisten, but not suffocate them.

If you want to go ape, you can place your blue claws in an ice-filled cooler. Place a piece of cloth or seaweed over the ice so the crabs don't rest against the ice but receive its cooling effects. Keep the cooler slightly ajar for air.

Another point to remember when trying to keep your catch alive is to make sure the crabs lay upright. Many crabs die simply because they are flipped upside down and have numerous other claws poking through their underbelly.

Besides flipping crabs rightside up, be sure to spread your crabs out between a few buckets rather than cramming fifty of them in one 5-gallon bucket while another empty bucket is flipped over with someone's keister occupying it.

If, for all your efforts, a crab or two die, discard these casualties. Either toss them over, mash them up for chum or save them for tog bait but don't eat them. Some say that keeping a freshly dead crab on ice will allow you to eat it without any ill effects, others say you should not eat a dead crab no matter what. Logically, it is not worth trying to eat one crab and risk food poisoning when you can catch many more. Though this may not sound like a conservationist approach, if you take time to care for the blue claws you catch, this dilemma will never confront you.

When and Where

If you are really itchin' to catch blue claws, they begin to become active once the water temperature goes over 50 degrees. This usually occurs around early May, and crabbers working the bay's shallow marshes typically put together the better catches. For the most part, crabbing does not really start to take off until around mid-June and then steadily gains ground through July into August. Though most crabbers hang up their handlines and traps after Labor Day, crabbing remains quite good into October. Though the raw number of crabs may be reduced by late September, the size of the crabs you are likely to catch is usually the biggest you'll see all year.

Essentially, almost any location within Barnegat Bay holds blue claws. Blue claws will be concentrated where there they can hide from predators plus enjoy a steady food supply. Salt marshes, creek and river mouths, estuaries, bridges, bulkheads, old pilings, eel grass flats, residential lagoons, and lazily moving back bay tidal pools are the prime areas that attract blue claws.

The aspect that basically separates good crabbing spots from great spots is whether one can successfully crab them when considering wind, boat traffic, current, and depth.

The wind often plays a major factor in crabbing because it causes anchored boats to swing and bob with the waves. The bay's backwater marshes offer considerable protection from the wind. Though these areas are typically shallow, allowing you to actually see your bait, don't let their shallowness fool you. A blue claw's dark brown shell camouflages it against a muddy bottom making specimens with an 8-inch width nearly invisible in 4-foot deep water.

Deep locations, typically with strong currents, make it difficult to keep a bait on the bottom unless one wants to overload the bait with lead. Even then, when retrieving the bait to the surface, the current will cause the bait to "plane" (i.e. ride horizontal along the surface) so it comes to the surface far down current from the crabber's position. The crabber literally has to drag the bait and crab along the surface to bring them within netting distance.

Most areas that offer great crabbing also boast light tackle fishing opportunities. Eels, snappers, oyster crackers, and blowfish, are the typical species that frequent good crabbing locations because they feed on blue claws. One time while crabbing near Gulf Point, my family and I accomplished the rare feet of catching a sheepshead porgy. The fish, which will prey on blue claws, measured roughly 12-inches long and its plump, vibrant appearance strongly suggested that it wasn't starving.

Upper Bay

As stated earlier, it would be hard to name an area within Barnegat Bay that did not host blue claw crabs.

When perusing the Bay Head area, the better spots seem to center

Did You Know?
Blue claw crabs are abundant within the Nile River and along the coasts of Israel and Lebanon.

Did You Know?
A scientific tagging study documented female blue claws moving 500 miles in 100 days.

around the Beaver Dam Creek and the Metedeconk River. For many, the bridge that spans across Beaver Dam Creek is a good place to catch blue claws. Try investigating this bridge at night or very early in the morning when the crabs are active and most boaters are not.

Just south of the Beaver Dam Creek flows the Metedeconk River. With its numerous feeder creeks, the Metedeconk is a prime area to crab especially during August. Crabbing is typically good from the Metedeconk's mouth west to Kingfisher Cove. When crabbing the Metedeconk, and the bay's other rivers and creeks, monitor the recent rainfall because blue claws are very sensitive to the water's clarity and salinity. If a monsoon-like rain dirties the water or makes it too fresh, the crabbing could be thrown off for a few days.

Crabbing hotspots that stick out south of the Metedeconk River are the Mantoloking Bridge, Kettle Creek, the Route 37 Bridge, and the Toms River. Both bridges really lend themselves to scraping and to scoop netting crabs at night. To many, the old pilings that lie near the Route 37 Bridge are good to drop handlines near and scrape. During August, it's possible in these areas to simple select any small patch of bottom, where you can anchor without fear of getting run over by boat traffic, and catch a good basketful of crabs.

The Toms River is a very popular crabbing area with crabbing activities centering around Good Luck Point, the Mathis Bridge, and the Island Heights area. Crabbers should investigate the many small creeks, such as the Dillon and Mill Creeks, that flow into the Toms River. Crabbing at the

mouths of these small creeks is generally productive.

South of the Toms River, Cedar Creek and Stouts Creek are two particular good crab locations whether crabbers target the creeks themselves, the shallow water at the creeks' mouths or the deeper water just east of these two creeks.

Near the bay's eastern shores, the flats attract crabs with Tices Shoals being one particular location to investigate. The better times to crab along the eastern side is during the off hours when the bay is quiet and still, and blue claws begin to stir. The flats' shallow water coupled with a powerful flashlight makes spotting blue claws pretty easy.

Lower Bay

Similar to the upper bay, the lower bay's western shores provide some splendid crabbing opportunities. Crabbers should check out the creeks, rivers, lagoons, marshes, and islands that dot the lower bay because these are the locations that typically hold the greatest blue claw populations.

Aside from the lower bay's western shores, Oyster Creek and Double Creek Channels provide good crabbing if crabbers investigate the shallows just outside the channels. Be sure to bring sinkers in the 4-ounce neighborhood if you plan to crab at the channels' eastern ends because when the tide begins to flow, you will have a tough time getting your bait to hold bottom. Also, you may want to try morning or dusk to avoid boat traffic.

Clam Island, which flanks Double Creek Channel, is usually a good

Did You Know?
Female blue claws will mate only once while males will mate several times.

crabbing area to check out.

Not far from Double Creek Channel lie several good crabbing grounds off Conklin Island, Long Beach Island's western shores and the Barnegat Township Public Docks, with the marshes around these areas particularly good for blue claws.

Barnegat Inlet also lends itself to crabbing, especially at night when the

Some of the turtles that I've seen back in the swamps taking my bait would call for one heck of a net if an average sized person could even wield such a device. When crabbing around the salt marshes remember to bring extra bait because a hungry snapping turtle can deprive you of your bait with a single snap!

However they're prepared, blue claws are a summertime treat like

boat traffic has ceased. Crabbing in the inlet's main is not a good idea, because of the thick boat traffic, but poking around the inlet's marshy banks, rock piles, and jetties is usually productive. A moving tide is an especially good time to look for blue claws because the inlet's strong current will buffet the crabs about making it much easier to net them.

When crabbing near or within marshes, such as the ones around Barnegat Township, Cedar Creek, the Toms River, or the Metedeconk River, snapping turtles will occasionally investigate the scene and make off with a few baits. Crabbers will occasionally catch smaller turtles with crab nets.

corn-on-the-cob. If you enjoy fishing, don't discard the shells and innards left over from a crab feast. Save the shells and juices in a bucket and use this "goulash" for chum. Weakfish and blowfish are two prime examples of fish that respond favorably to chum derived from blue claw crabs.

Barnegat Bay's backwater marshes, creeks and rivers are prime locations to catch blue claw crabs. Be sure not to raise a ruckus when poking around these quiet and relatively shallow locations for crabs (center).

Conclusions

As you can see, anglers could spend years fishing Barnegat Bay and never feel bored or complacent because of all the fish species and fish holding locations the bay possesses. Part of the thrill of fishing Barnegat bay is to be able to enjoy parts of its unspoiled landscape. Seeing the light, morning mist hang over a low marsh that stretches for miles like a green sea gives you the sense of what this land must have looked like long ago.

In parting, it's my sincere hope that everyone, while demonstrating their angling techniques, demonstrates conservation through some common sense by keeping only what can be sensibly used.

Maybe you and I will rub elbows some day while on the bay . . . . . .

About the Author

From poking around for brook trout in backwoods streams to battling cod and the arctic-like conditions of offshore winter wreck trips, Rich Henderson enjoys fishing for a wide variety of fish species. Rich has been fishing and crabbing at Barnegat Bay for over 22 years and considers it to be his home turf.

A member of the International Game Fish Association and Trout Unlimited while being a former member of the American Littoral Society, Rich likes to further conservation efforts towards the national and global fisheries. Rich has also worked with the National Oceanic and Atmospheric Administration (NOAA) in its legislative efforts towards the national fisheries.

An avid writer, Rich regularly contributes to The Fisherman magazine and to Greater Media Newspapers.